Commercial Breeding and Seed Production of Vegetable Crops

NIPA® GENX ELECTRONIC RESOURCES & SOLUTIONS P. LTD.
New Delhi-110 034

About the Editors

Dr. E Sreenivasa Rao, Principal Scientist at the Division of Vegetable Crops ICAR-IIHR, Bangalore has completed his PhD from IARI New Delhi and Post Doctoral from World Vegetable Center Taiwan. He has more than 24 years of experience in the field of vegetable crop improvement and has contributed significantly in the area of resistance breeding of cucurbits, heterosis breeding of onion & okra and salinity tolerance in tomato. He has published more than 55 research articles, developed 13 varieties and F1 hybrids in onion, carrot watermelon, muskmelon, ridge gourd and okra which are well accepted by farmers and seed is in demand. He has carried out pioneering work in developing synthetic varieties of onion and disease resistant root stocks for watermelon grafting. Three disease resistant genetic stocks in watermelon and ridge gourd have been registered with NBPGR. He has identified novel alleles of *Dreb1* gene for salinity tolerance in tomato, *Clfs1* gene for fruit shape in watermelon and *Cmph* gene for fruit acidity trait of muskmelon. As Advisor to the Government of Telangana, he is now involved in the Oilpalm plantation drive in the state with a target of 4 lakh ha in the next three years. He has also initiated a carbon finance program for Agro-forestry plantations in the state.

Dr. B.Varalakshmi, Principal Scientist at the Division of Vegetable Crops ICAR-IIHR, Bengaluru has completed PhD in Horticulture from the then AP Agricultural University, Hyderabad, Andhra Pradesh with specialization in Vegetable breeding. She has 32 years of vast experience in breeding of different vegetable crops viz., ridge gourd, bitter gourd, tropical cauliflower and leafy vegetables. So far, she has developed 9 varieties and a hybrid in these crops *viz.,* Arka Prasan, (notified for Karnataka), Arka Vikram (F_1 notified for Zone 4 and 8) (Ridge gourd), Arka Vimal, Arka Spoorthi (Cauliflower) Arka Suguna, Arka Samraksha, Arka Varna, Arka Arunima, (Amaranth, first 3 notified for Karnataka), Arka Anupama (Palak notified for Karnataka) and Arka Isha (Coriander). She is responsible for maintaining the germplasm and seed production of these released varieties. She isolated unique CGMS line of ridge gourd and this trait is being transferred to desired back grounds to be used in hybrids seed production. She also developed gynoecious lines in bitter gourd. She reported for the first time the powdery mildew resistance in bitter gourd line, IIHR-144-1, which has been registered with NBPGR and also identified the SSR marker linked to resistance. She has guided 5 PhD and 6 M.Sc. students. She has 117 various publications to her credit including 46 research articles in the peer reviewed international and national journals.

Dr. Naresh Ponnam, did B.Sc (Horticulture) from ANGRAU, Hyderabad M.Sc (Horticulture) from University of Agricultural Sciences, Bengaluru and Ph.D (Horticulture) from University of Horticultural Sciences, Bagalkot. He holds considerable experience in chilli breeding, germplasm maintenance hybridization and handling of the populations and forwarding generations through Marker assisted selection. He has undergone SERB- Overseas Post-doctoral fellowship at World Vegetable Centre, Taiwan on Genotyping-by-sequencing, and SNP markers. Presently working as Senior Scientist at Division of Vegetable Crops, ICAR-Indian Institute of Horticultural Research, Bengaluru. He handled external projects mainly on chilli breeding programs funded by DST, NASF, MIDH and RKVY etc. He has published 36 research articles in national and international peer-reviewed journals and four book chapters and several popular articles.

Dr. Saheb Pal, obtained his B.Sc. (Horticulture) degree in 2014 from BCKV Mohanpur, West Bengal; M.Sc. (Horticulture) Vegetable Science degree in 2016 from Dr YSPUHF, Solan, Himachal Pradesh and PhD (Vegetable Science) degree from ICAR-IARI, New Delhi (Outreach Campus: ICAR-IIHR, Bengaluru). He is the recipient of University Gold Medals in both B.Sc. and M.Sc. degree programmes. He is also the recipient of the prestigious "Best Student of the Year 2020" award at ICAR-IARI, New Delhi and "Best Student of the Year 2018 and 2021" at ICAR-IIHR, Bengaluru. He secured AIR-1 in the ICAR-ASRB-ARS-2021 Examination in the Discipline of Vegetable Science. Presently he is working as a Scientist (Vegetable Science) at the ICAR-Indian Agricultural Research Institute, Gauria Karma, Barhi, Hazaribagh Jharkhand-825405 and presently working on genetic improvement of okra.

Commercial Breeding and Seed Production of Vegetable Crops

E Sreenivasa Rao
Principal Scientist
Division of Vegetable Crops
ICAR-IIHR, Bangalore, Karnataka

B.Varalakshmi
Principal Scientist
Division of Vegetable Crops
ICAR-IIHR, Bengaluru, Karnataka

Naresh Ponnam
Senior Scientist
Division of Vegetable Crops
ICAR-Indian Institute of Horticultural Research
Bengaluru, Karnataka

Saheb Pal
Scientist (Vegetable Science)
ICAR-Indian Agricultural Research Institute
Gauria Karma, Barhi, Hazaribagh, Jharkhand

NIPA® GENX ELECTRONIC RESOURCES & SOLUTIONS P. LTD.
New Delhi-110 034

NIPA® GENX ELECTRONIC RESOURCES & SOLUTIONS P. LTD.

101,103, Vikas Surya Plaza, CU Block
L.S.C. Market, Pitam Pura, New Delhi-110 034
Ph : +91-11-43860225, Mob.: +91 9717133558, 9540816132
E-mail: newindiapublishingagency@gmail.com
Website: www.nipaersources.com

Print ISBN: 978-93-58871-35-7
ebook ISBN: 978-93-58879-52-0

Composed and Designed by NIPA®.

Preface

The importance of seed for human prosperity has been mentioned in Rigveda ***'as Subeejam Sukshetre Jayate Sampadyate'*** meaning good seed in field will win and prosper. Availability of quality seed is a key and essential requisite for farming. Commercial seed production is currently a multibillion-dollar business globally. Even in our country, vegetable seed industry is growing at a very fast pace annually. Several government agencies and private companies are actively involved in the seed production and marketing. The Indian vegetable seed industry is projected at CAGR of 7-8% for the forecast period between 2020-2025. Vegetable seed market in India is valued at 655.2 million US $ in 2022 and is expected to rise to 931.76 million US $ by 2030. Thus, vegetable seed industry has positive influence on Indian economy in terms of income and employment generation and earning foreign exchange in international market. Further, there is also a need to involve and encourage rural youth in this business model through efficient training and dissemination of seed production techniques. Unlike field crops, vegetable crop seeds are low volume and high value; also require specific interventions during pre- and post-harvest stages of seed production. Vast diversity in pollination behavior of vegetable crops requires specific seed production and post-harvest production techniques for different crops. Further standards are specifically defined to ensure physical and genetic purity to meet high quality. Therefore, an effort is being made to compile and present these techniques for a practitioner of vegetable seed production. This book contains thirteen chapters on commercial breeding and seed production techniques of major vegetable crops like tomato chilli, capsicum, brinjal, melons, legume vegetables, onion and okra. This book has highlighted the commercial breeding, pollination behavior, scientific seed production techniques, seed processing techniques and seed standards required in each crop.

We hope this book will help seed entrepreneurs, breeders, students, teachers, researchers and scientists in the field of horticulture. We are extremely grateful to the contributors in compiling the available information and their personal expertise in the form of a book for practical understanding of commercial breeding and seed production of vegetable crops.

Editors

Contents

1

Tomato

H.A. Bhargavi and H.C. Prasanna

Division of Vegetable Crops, ICAR-Indian Institute of Horticultural Research Bengaluru, Karnatka

Abstract

Tomato (Solanum lycopersicum L.) is one of the most commercially important vegetable crops globally, with a significant demand for high-yielding hybrids and improved open-pollinated varieties. Efficient seed production methodologies play a crucial role in ensuring the availability of genetically pure and high-quality seeds for commercial cultivation. Traditional hand emasculation and pollination remain widely used despite being labor-intensive, given their economic feasibility due to high seed yield per fruit. Hybrid seed production is further refined through mechanized pollen collection and controlled pollination techniques, enhancing efficiency. Seed quality management, including rigorous isolation protocols, field inspections, and breeder and foundation seed production standards, ensures genetic purity and viability. Advances in molecular breeding and the application of marker-assisted selection (MAS) have enhanced the efficiency of parental line development and hybrid seed production. This chapter provides an in-depth review of tomato seed production methodologies, covering botanical characteristics, genetic resources, and breeding objectives aimed at improving disease resistance, abiotic stress tolerance, and fruit quality traits. It further discusses varietal advancements in the public and private sectors, highlighting elite hybrids and open-pollinated cultivars tailored for different agro-climatic conditions. Additionally, the package of practices for seed production in tomato underscores the importance of isolation, field inspections, seed extraction methods, processing techniques, and quality assurance standards to maintain genetic purity and seed viability.

Keywords: *Tomato, hybrids, genetic purity, heterosis, yield*

Introduction

Tomato (*Solanum lycopersicum* L.; 2n=24) is one of the most widely cultivated and economically significant vegetable crops globally, belonging to the

Solanaceae family. The top tomato-producing countries include China, India, Turkey, United States, Egypt and Italy (FAOSTAT, 2023). It is predominantly a self-pollinated species, but hybrid breeding has been extensively utilized to enhance yield, stress tolerance, and fruit quality. The genus Solanum includes several wild relatives of tomato, with *Solanum pimpinellifolium* recognized as its closest wild progenitor (Peralta *et al.*, 2006). A total of 12 wild species have been identified in the *Solanum* section *Lycopersicon*, many of which serve as valuable genetic resources for breeding programs. It is highly valued as a protective food due to its rich composition essential nutrients, including vitamins (A, C, and E), minerals, and bioactive compounds such as lycopene, a potent antioxidant with significant health benefits (Collins *et al.*, 2022). Tomato is a staple dietary component in many countries, including the United States, Italy, China, India, Spain, Mexico, and Turkey (Valpuesta, 2002). It holds an integral place in global cuisines due to its versatility, flavor, and nutritional value. The rising demand for high-yielding, disease-resistant, and climate-resilient varieties has led to continuous advancements in commercial breeding and seed production technologies.

Tomato breeding has traditionally relied on selection and hybridization to improve traits such as yield, fruit quality, disease resistance, and adaptability. However, modern breeding approaches, including marker-assisted selection (MAS), genomic selection, and genome editing (CRISPR-Cas9), have accelerated genetic improvement. Hybrid breeding has gained prominence due to the heterotic advantage of F_1 hybrids, which exhibit superior agronomic traits over open-pollinated varieties (OPVs). Furthermore, Tomato production is challenged by various biotic and abiotic stresses that negatively impact plant growth, significantly reduce yield, and compromise fruit quality (Liedl *et al.*, 2013). Therefore, breeding efforts prioritize earliness, yield improvement, fruit quality enhancement, and resistance to key biotic (late blight, early blight, whitefly) and abiotic (drought, salinity, heat) stresses. The crop is cultivated in both open-field and protected environments, with the latter gaining prominence for year-round production and higher yield potential. Hybrid seed production in tomato traditionally relies on manual emasculation and pollination, making the process labor-intensive and costly due to the small flower size and delicate floral structures. While the GMS and CGMS systems in tomato have potential, their widespread commercial application is currently limited due to biological and technical constraints. Seed processing and quality management play a critical role in ensuring seed viability, vigor, and genetic integrity. Key steps include seed extraction, fermentation, drying, grading, and treatment to prevent seed-borne diseases and improve storage life. Quality control measures such as germination tests, moisture content evaluation, and adherence to certification standards (ISTA, OECD) are essential for producing seeds that meet market and regulatory requirements.

Botany and Wild Species

Tomato (*Solanum lycopersicum*), a member of the Solanaceae family, which comprises over 3,000 species adapted to a wide range of habitats (Knapp 2002). The tomato plant originated in the Andean region of South America where wild species continue to thrive in their native habitat. In addition to domesticated species, the tomato gene pool comprises 12 wild relatives that serve as valuable genetic resources for breeding programs. The 12 wild species include *Solanum pimpinellifolium, S. arcanum, S. cheesmaniae, S. chmielewskii, S. chilense, S. galapagense, S. corneliomulleri, S. huaylasense, S. habrochaites, S. peruvianum, S. neorickii*, and *S. pennellii* (Peralta *et al.* 2008). These wild species exhibit a wide range of traits, including resistance to biotic and abiotic stresses, as well as variations in fruit morphology and biochemical composition. Understanding their genetic diversity is crucial for improving cultivated tomato varieties.

It is a self-pollinated, herbaceous plant with pinnately compound non-spiny leaves, bright yellow flowers, and fleshy berries that vary in shape, size, and color. Its domestication traces back to the Andean region of South America with *Solanum pimpinellifolium* considered its closest wild ancestor (Li *et al.*, 2022). The domesticated species evolved through selection for larger fruit size, improved yield, and enhanced flavor, with key genetic modifications affecting fruit weight (*FW2.2*), shape (*SUN*), and locule number (*LC, FAS*) (Knaap *et al.*, 2014; Chu *et al.*, 2019). Modern breeding integrates genes from wild relatives like *S. peruvianum, S. habrochaites*, and *S. pennellii* to enhance disease resistance, stress tolerance, and shelf life (Panthee and Chen., 2010). Today, cultivated tomatoes are classified into determinate and indeterminate growth types, catering to diverse agricultural and culinary needs.

Genetic Resources

The success of any crop breeding program relies mainly on the availability of genetic resources. Approximately 40,365 tomato accessions are preserved in global gene bank collections (https://www.genesys-pgr.org). NBPGR India, maintains a diverse tomato germplasm collection, including landraces cultivars, and wild relatives, which maintains 2931 accessions (https://nbpgrorg.in/nbpgr2023/genebank-status-2/). The C.M. Rick Tomato Genetics Resource Center (TGRC) conserves wild tomato relatives, monogenic mutants, and diverse genetic stocks. The monogenic collection includes 1,058 accessions (630 loci) with mutations affecting growth, development, and disease resistance.

Miscellaneous stocks comprise multiple marker combinations, cytogenetic variants (trisomics, tetraploids, translocations), cultivars landraces, and prebred lines (introgression, backcross inbred, recombinant inbred, and monosomic alien addition lines). The latest update includes 1,203 accessions from 17 wild *Solanum* species and 2,201 miscellaneous genetic stocks. These resources support breeding for yield, quality, and stress resistance. Characterized using agronomic and molecular tools, the collection ensures genetic diversity for future crop improvement.

Floral Biology and Pollination Mechanisms

Tomato plants exhibit extra axillary cymes characterized by dichotomous or polychotomous branching for inflorescences, which emerge laterally from the stem, opposite to the leaf. These inflorescences may terminate or continue with vegetative shoots. Bright yellow flowers are produced, with the first flower bud in the initial inflorescence originating from the terminal meristem of the main stem. Subsequent buds develop laterally on the pedicels of preceding flowers. Flowers are arranged alternately on both sides of the inflorescence peduncle. Initially, the peduncle displays a helical structure unfurls as the first flower opens and grows into a fruit. The flowers are characterized as pentamerous, bisexual, regular, complete, ebracteate, and hypogynous. The pistil typically comprises two to several carpels. Anthers are connate and situated within the corolla's throat, with introrsely longitudinal dehiscence occurring either at or shortly after the corolla's opening. Because the style is shorter than the anther, self-pollination occurs frequently.

Tomato flowers typically begin floral anthesis around 6 a.m. and continue until 11 a.m., the peak time of anther opening is between 8 to 11 a.m., influenced by factors like sunshine, humidity, and temperature. Pollen remains viable for 2 to 5 days in temperatures ranging from 18° to 25°C. Meanwhile, the stigma is receptive for 16 to 18 hours prior to anthesis and remains so for up to 6 days afterward, just before the flower withering. Again, this extended period of stigma receptivity allows for controlled pollination, supported by the extended viability of pollen, which remains viable for 2 to 5 days at temperatures between 18° to 25°C and for up to six months at 5°C in a desiccator. Optimal fertilization and fruit set occur when pollination aligns with anthesis, typically 24-50 hours post-pollination.

Tomato Breeding: Domestication to Modern Cultivars

Numerous tomato cultivars in various colours and uses have emerged as landraces through domestication and early breeding. These open-pollinated cultivars were propagated by farmers, who saved seeds from fruits for future planting. Meanwhile, the emergence of public institutions and private companies led to a shift from open-pollinated cultivars to hybrid development in commercial breeding. Tomato breeding has resulted in significant diversity in plant architecture, growth habit, and fruit characteristics, including shape, size, color, and flavor. Presently, most commercially available tomato cultivars are classified into two main categories: fresh market (FM) and processing (PROC) types. Traditional tomato breeding relies on gene transfer from wild relatives into modern cultivars. Techniques such as mass selection, pedigree breeding, hybridization, and molecular marker-assisted selection are employed to develop linkage maps and enhance desirable traits. The advent of DNA-based molecular markers like RFLPs and AFLPs has enabled the development of complete genetic maps in tomatoes (Tanksley *et al.*, 1992; Haanstra *et al.*, 1999).

Genetic mapping has proven valuable in enhancing fruit weight and fruit lycopene content (Gonda *et al.*, 2019), diversifying fruit colors such as red, pink, yellow, green, and purple (Ballester *et al.*, 2010), late blight resistance (Jia, 2019), early blight (Adhikari *et al.*, 2023) and extending fruit shelf life (Vrebalov *et al.*, 2002). Significant advancements in tomato breeding have been made by integrating conventional methods with marker-assisted selection (MAS). Tomato was the first crop in which molecular markers were applied for genetic mapping and breeding advancements (Tanksley and Rick, 1980). Marker-assisted selection (MAS) in tomato breeding dates back to the 1930s, making it one of the earliest crops to adopt this approach. It has been widely utilized for enhancing morphological, disease resistance, and physiological traits. Marker-assisted selection (MAS) has been widely applied in breeding for disease resistance Fusarium wilt, late blight, leaf mold, bacterial speck, tomato mosaic virus and Verticillium wilt, fruit color, ripening genes, carotenoid content, extended field storage and jointless pedicel (Osei *et al.*, 2018). It is regularly used in the seed industry for traits like resistance to Fusarium wilt (races 1–3), late blight (*Ph-3*), bacterial spot (*Rx3*, *Rx4*), Verticillium wilt (race 1), tomato yellow leaf curl virus (*Ty1–Ty4*), tomato spotted wilt virus (*Sw5*), and root-knot nematode (*Mi*) (Tiwari *et al.*, 2022). Recently, high-throughput SNP genotyping has gained traction, enabling more precise tomato breeding.

The primary objectives of tomato breeding include enhancing fruit yield and quality, developing resistance to biotic stresses, and improving tolerance to abiotic factors.

Fruit Yield and Heterosis

It's a complex trait determined by the number of fruits per plant, as well as their size and weight. Breeding for fruit yield in tomato focuses on improving fruit number, size, weight, and overall productivity through conventional, molecular, and genomic approaches. Quantitative trait loci (QTLs) associated with yield have been mapped and utilized in marker-assisted selection (MAS) to develop high-yielding cultivars (Ning *et al.*, 2023). Hybrid breeding has significantly enhanced yield potential by exploiting heterosis, while genomic selection (GS) and CRISPR-based gene editing are emerging as precise tools for trait improvement. Wild tomato species such as *Solanum pennellii* and *S. habrochaites* have been valuable sources of genes for enhanced yield traits. Additionally, optimizing plant architecture, including determinate growth habit and improved source-sink balance, has contributed to increased productivity. However, challenges like climate variability, resource limitations, and disease pressures require continuous breeding efforts to develop stable, high-yielding tomato cultivars.

Fruit Quality

Breeding for fruit quality in tomato focuses on multiple traits, including visual appeal, nutritional content, shelf life, and flavor. Fresh-market and processing varieties require distinct quality attributes. Fresh-market tomatoes prioritize extended shelf life, ideal shape, and vibrant color (Fentik, 2017), whereas processing varieties must meet specific standards such as high total soluble solids (TSS >5.5° Brix), high lycopene levels (>14 mg/100 g fresh weight) with a high colour value of (>2), optimal acidity (0.35 to 0.4%), minimum sugar acid ratio (15:1) and high vitamin C content (>25 mg/100 g). Additional factors like low pH (<4.4), firmness (>4), pericarp thickness (>0.4 cm), and viscosity (12–14 Bostwick cm/30s) are critical for processing tomatoes, along with traits such as uniform red color, jointless pedicel for mechanical harvesting, and suitable juice yield (>70%) (Tigist *et al.*, 2013). The early tomato breeding programs in India, particularly in the 1980s, resulted in the development of several open-pollinated (OP) varieties suitable for processing. Notable releases include Roma (NBPGR, New Delhi), Arka Ashish and Arka Ahuti (IIHR, Bengaluru), Punjab Chhuhara (PAU, Ludhiana), and Pusa Gaurav (IARI, New Delhi).

Ripening plays a crucial role in determining fruit quality as it affects color, flavor, and storage potential. Genes such as polygalacturonase and ethylene synthase regulate ripening and have been successfully cloned (Miller *et al.*, 2002). Lycopene, a key antioxidant responsible for the red color of tomatoes, accumulates significantly during ripening up to 500 fold and provides substantial health benefits, including a reduced risk of certain cancers. High-lycopene tomato varieties have been introduced as premium products in response to consumer demand for nutritionally enhanced tomatoes. Flavor is a complex trait influenced by the interaction of sugars, acids, and volatile compounds. Approximately 30 volatile compounds contribute to tomato flavor, which can be partially predicted by measuring acidity and soluble solids content using the refractive index (Tieman *et al.*, 2006). Breeding efforts focus on enhancing sugar and acid balance while optimizing key volatiles to improve taste. Despite challenges in precisely predicting flavor, genetic manipulation and marker-assisted selection have enabled significant advancements. Modern breeding strategies continue to refine fruit quality traits, ensuring tomatoes meet both consumer preferences and industry standards.

Resistance to Biotic Stresses

Tomato (*Solanum lycopersicum*) is highly susceptible to a range of bacterial, fungal, viral, and nematode diseases, which pose significant challenges to its cultivation. Breeding for resistance to biotic stresses has been a primary focus in tomato improvement programs, with efforts centered on the identification and introgression of resistance genes from wild relatives into cultivated varieties. Host resistance genes have been successfully leveraged to combat major diseases, including Fusarium wilt (*Fusarium oxysporum*), tomato leaf mold (*Cladosporium fulvum*), Verticillium wilt (*Verticillium albo-atrum*), late blight (*Phytophthora infestans*), bacterial speck (*Pseudomonas syringae*), bacterial spot (*Xanthomonas* spp.), tomato yellow leaf curl virus (TYLCV), tomato mosaic virus (ToMV), tomato spotted wilt virus (TSWV), and root-knot nematodes (*Meloidogyne* spp.). The introgression of resistance alleles from wild tomato species has played a crucial role in developing resistant cultivars. For instance, *Solanum pimpinellifolium* has been utilized as a genetic source for developing resistance against *Cladosporium fulvum*, the causal agent of tomato leaf mold (Bai and Lindhout, 2007).

Resistance to Tomato Mosaic Virus (ToMV) has been identified in *Solanum peruvianum*, where three major resistance genes *Tm1, Tm2*, and *Tm2a* have been reported. Among these, *Tm2a* provides broad-spectrum resistance and is widely incorporated into breeding programs (Panthee *et al.*, 2013). Similarly, resistance to begomoviruses such as TYLCV has been introgressed from wild tomato species, with six resistance loci Ty-1, Ty-2, Ty-3, Ty-4, Ty-5, and Ty-6 reported to confer resistance (Gill *et al.*, 2019; Al-Saikhan *et al.*, 2020). Late blight, caused by *Phytophthora infestans*, remains one of the most destructive diseases affecting tomato and potato crops. Five resistance genes have been identified and incorporated into breeding programs, with *S. pimpinellifolium* serving as a valuable source of resistance. The *Ph-1* gene, first identified in *S. pimpinellifolium*, was mapped to the long arm of chromosome 7 and is widely utilized for late blight resistance breeding. Bacterial wilt resistance sources have also been identified in *S. lycopersicum var. cerasiforme* and *S. pimpinellifolium*, providing valuable genetic material for breeding programs (Aslam *et al.*, 2017). Tomato is also a major host for root-knot nematodes (*Meloidogyne* spp.), leading to significant yield losses. The *Mi-1* gene is the most commercially successful gene used in breeding to develop resistant cultivars, providing protection against *M. incognita, M. arenaria*, and *M. javanica*, though it is ineffective against *M. hapla*. The *Mi-2* gene, in contrast, has been reported to confer resistance to four nematode species (El-Sappah *et al.*, 2019). Overall, the strategic introgression of resistance genes from wild tomato species continues to be a key approach in breeding for biotic stress resistance. Advances in molecular breeding techniques, including marker-assisted selection (MAS) and genomic selection, are further enhancing the efficiency of developing resistant tomato cultivars.

Tolerance to Abiotic Stresses

Tomato (*Solanum lycopersicum*) production is significantly impacted by abiotic stresses such as high temperature, salinity, and drought. To mitigate these challenges, breeding programs have focused on identifying and introgressing tolerance traits from wild tomato relatives, utilizing molecular tools such as quantitative trait loci (QTL) mapping, marker-assisted selection (MAS), and genomic selection (GS). Advances in molecular marker technology have facilitated the identification of key genomic regions associated with abiotic stress tolerance, enabling more efficient breeding strategies. Salinity tolerance in tomato has been extensively studied, with wild species such as *S. pimpinellifolium, S. cheesmaniae, S. peruvianum*, and *S. pennellii* serving as valuable genetic resources. QTL analysis has pinpointed genomic regions linked to salt tolerance, with MAS proving effective in enhancing salinity resilience. Notably, studies by Foolad *et al.* (2001) demonstrated that interspecific variation during the vegetative stage could be exploited to improve salt tolerance. Molecular markers associated with rapid seed germination under salinity stress have been identified in wild tomato species, contributing to the development of salt-tolerant cultivars.

Heat tolerance research in tomato has progressed through the identification of heat-responsive genes, QTLs, and molecular markers. Key molecular components, including heat shock proteins (HSPs), heat shock transcription factors (HSFs), and hormonal pathways, have been implicated in thermotolerance (Liu *et al.*, 2023). The integration of MAS and GS has accelerated breeding for heat-resilient cultivars, although challenges remain in developing high-yielding varieties with stable performance across diverse environments. Emerging technologies such as CRISPR-based gene editing and multi-omics approaches offer new avenues for improving heat tolerance. Drought tolerance in tomato has been explored through physiological, genetic, and molecular studies. Wild relatives like *S. pennellii* and *S. pimpinellifolium* have been utilized for their drought resilience traits, contributing to the identification of drought-responsive genes and QTLs. Transcriptomic and metabolomic studies have highlighted key mechanisms of drought adaptation, including osmoprotectant accumulation, antioxidant enzyme activity, and hormonal regulation. Despite these advances, challenges persist in developing high-yielding, drought-tolerant cultivars with consistent performance under variable environmental conditions. Genomic selection and gene-editing tools such as CRISPR hold great promise for future improvements in drought resilience. Overall, breeding for abiotic stress tolerance in tomato has made significant strides, driven by molecular advancements and the exploitation of wild genetic resources.

Varietal Achievements

Table 1: Mutant varieties and hybrids developed from various public institutions from India

Varieties	Breeding methods	Special features	Source
Mutant Variety			
CO-3 (Marutham)	Mutant of CO-1	Determinate and cluster bearing type	TNAU, Coimbatore
S-12	X-ray mutant of Sioux	Round fruited and high yielding variety	PAU, Ludhiana
PKM-1	Mutant of Annanji	Green flesh type, Long distance transport	TNAU, Coimbatore
Pusa Lal Meeruti	Gamma ray mutant of Meeruti	Uniform fruit ripening type and disease resistant variety	IARI, New Delhi
Varieties and Hybrids			
Arka Vishal	IHR 837 × IHR 932	Tolerant to cracking and suitable for fresh market	IIHR, Bangalore
Arka Vardan	IHR 550-3 × IHR 932	Indeterminate and tolerant to fruit cracking	IIHR, Bangalore
Arka Shreshta	15 SBSB × IIHR 1614	Resistant to bacterial wilt	IIHR, Bangalore
Arka Samrat	IIHR-2835 x IIHR-2832	Resistant to bacterial wilt, ToLCV and early blight	IIHR, Bangalore
Arka Rakshak	IIHR-2834 x IIHR-2833	Resistant to early blight, bacterial wilt and ToLCV	IIHR, Bangalore
Arka Abhijit	15 SBSB x IIHR 1334	Resistant to bacterial wilt	IIHR, Bangalore
Pusa Divya	Long style × Roma	Developed using male sterile line, Antherless mutant	IARI, New Delhi

Pusa Hybrid-1	Pusa Sheetal × Chikoo	Fruit set at high night temperature	IARI, New Delhi
Pusa Hybrid-2	Pusa-120 × Pusa Gaurav	Highly tolerant to root knot nematode	IARI, New Delhi
Pusa Hybrid-4	Pusa-120 × Chikoo	Field resistance to root knot nematode	IARI, New Delhi

Table 2: Improved cultivars of tomato developed by public sector in India

Varieties	**Breeding methods**	**Special features**
IARI Varieties		
Pusa Red Plum	*S. lycopersicum* × *S. pimpinellifolium*	Interspecific hybrid, Rich in Vitamin-C
Pusa Ruby	Sioux × Improved Meeruti	Most famous variety of tomato
Pusa Early Dwarf	Improved Meeruti × Red Cloud	Early ripening determinate variety
Pusa Rohini	-	Extremely well-suited for extended transportation and processing.
Pusa-120	-	Resistant to root-knot nematode
Pusa Gaurav	-	Ideal for extended transportation and processing.
Pusa Uphar	-	Tolerant to fruit borer
Pusa Sheetal	-	Early and determinate type
Pusa Gaurav	-	Suitable for processing and long distance
IIHR Varieties		
Arka Vikas	Selection from Tip Top	Suitable for fresh market, rainfed variety
Arka Alok (BWR-5)	-	Resistant to bacterial wilt
Arka Abha (BWR-1)	-	Resistant to bacterial wilt
Arka Abhijit	-	Highly resistant to bacterial wilt
Arka Ashish	-	Fruit cracking and powdery mildew tolerant
Arka Saurabh	Selection from V-685	Suitable for fresh and long transport
Arka Ahuti	-	Suitable for processing
Arka Meghali	Arka Vikas × IHR-554	Rainfed variety
IIVR, Varanasi		
Kashi Vishesh	*S. habrochaites* B'6013'	Resistant to TLCV
Kashi Amrit	-	-
Kashi Hemant	-	-
Kashi Sharad	-	-
Kashi Anupam	-	-
Other Varieties		
Hisar Lalit	-	Resistance to rook knot nematode

Hisar Anmol	*Hisar Arun* × *S. hirsutum f. glabratum*	Resistance to tomato leaf curl virus
Punjab Chhuhara	EC-55005 × Punjab Tropic	Suitable for long transport
CO-1	Selection from "Kalyanpur"	Semi-determinate
CO-2	Introduction-"Russia"	Indeterminate
Paiyur-1	Pusa Ruby × CO-3	Rainfed variety
MPKV, Rahuri		
Phule Raja	-	-
Dhanashree	-	-
Dr.B.S.Konkan Krishi Vidyapeeth, Dapoli		
Sonali	VC 48-1 × Tamu chico III	Bacterial wilt resistant

Hybrid Tomato Seeds can be Produced Through Three Methods

1. Hand emasculation and hand pollination.
2. Utilizing male sterility combined with hand pollination
3. Utilizing male sterility with natural pollination.

Hand Emasculation and Hand Pollination

Although labour-intensive, this approach remains economically feasible for crops such as tomatoes because of the substantial seed yield per fruit and the minimal seed rate per acre, which can be as low as 50 grams. Early advocates of heterosis breeding, like East and Heyes, underscored the pragmatic benefits of tomato hybrids, emphasizing the simplicity of crossbreeding in this Solanaceae plant. Diligent supervision of crop development is crucial, especially concerning irrigation and fertilization techniques, as moisture and nutrient deficiencies can result in diminished fruit and seed production. Seed yield fluctuates based on plant and fruit characteristics, spanning from 85-125 kg/ha for compact plants bearing blocky fruit, to 130-160 kg/ha for compact plants with round fruit, to 145-180 kg/ha for medium-large plants with round fruit, and to 140-180 kg/ha for indeterminate plants bearing round fruits. Seed companies have pledged their support to research programs aimed at hybrid tomato variety development and are also financing research initiatives focused on hybrid seed production.

Emasculation is usually carried out in the afternoon, ideally when the corolla petals have recently opened and are angled up to 45° from the flower axis, ensuring no self-pollination has occurred at this stage. Emasculated flowers are then pollinated the next morning since the stigma is not receptive on the day of emasculation (Figure 1). It is crucial to avoid delaying pollination to prevent the stigma surface from drying out, particularly in dry and hot windy conditions. Mechanized methods are utilized for pollen collection and its application to the stigma in large-scale hybrid seed production. In the case of tomatoes, it has been noted that around 51 minutes are required to produce 4000 hybrid seeds, and the use of pollen sterility, whether it is pollen sterility, positional sterility, or functional sterility, can notably decrease this timeframe. Emasculation contributes to approximately 40 percent of the overall labour in tomato hybrid seed production.

Figure 1: Emasculation and pollination in tomato: **A.** Selection of the flower at right stage; **B.** Removal of anthers; **C.** Pollen collection from male parent; **D.** Application of pollen on emasculated flower; **E.** Tagging of pollinated flower

Utilization male sterility and utility in F_1 hybrid seed production

Tomato male sterility is classified into four types, each controlled by a single recessive gene (Table 3). The Stamen less type results in deformed fruit.

Table 3. Characters of identified male sterile mutants in tomato breeding system.

Mutant	**Characters**	**Inheritance**	**Governing genes**
Pollen non functional	Pollen non functional	Monogeneic recessive	recessive
Stamenless /Stamens absent	Stamens less	Monogenic recessive (ms)	*"sl"*
Positional sterility (Ps)	Exerted Stigma	Monogenic recessive (ms)	*"ps"*
Functional sterility (Fs)	Anthers do not helps to Dehisce pollen	Monogenic recessive (ms)	*"ps-2"*

In F_1 hybrid production, positional sterility lacks consistency, whereas pollen abortive and functional sterilities are commonly employed. Pollen abortive sterility is maintained through backcrossing (ms ms × Ms ms), while functional male sterility is maintained through manual selfing. The efficiency of hybrid seed production using pollen abortive male sterile lines is improved by integrating morphological markers like the anthocyanin less gene 'aa,' which is associated with ms1035. This association facilitates the identification of male sterile plants amidst mixed populations of fertile and sterile plants in the nursery. As a result, only male sterile plants are transferred to hybrid seed production plots, reducing the need for space and labour. However, the efficacy of this approach hinges on the strength of the linkage between male sterility and anthocyanin less.

The effective use of male sterility genes depends on the availability of the style for cross-pollination. Disrupting the anther cone to expose the stigma for pollination undermines the advantage of male sterility if the style remains inaccessible. Combining male sterility with exerted stigma and the anthocyanin less gene presents inherent challenges, including the occurrence of 2-3% reciprocal recombinants (fertile plants lacking anthocyanin), low seed yield due to inadequate stigma receptivity, and the loss of exerted style in 20-60% of flowers, influenced by genotype and environmental factors. The exerted style trait exhibits greater stability in small-fruited genotypes than large-fruited ones.

Seed Production Specifications

Isolation: Despite their predominantly self-pollinated nature, tomato seed crops necessitate a 50m distance from different tomato varieties/hybrids or non-conforming varieties to maintain varietal purity in accordance with seed certification laws. This precautionary measure is implemented to prevent any potential outbreeding.

Land preparation: Tomato seed crop fields should not have any volunteer plants; thus, planting them in fields previously used for crops other than tomatoes is recommended. Seed certification standards do not specify specific requirements regarding the preceding crop.

Field Inspection: A key step in tomato seed production is conducting a minimum of three inspections throughout the crop season. These inspections play a vital role in identifying and removing off-type plants, thereby ensuring the authenticity of the seed crop. The first inspection, conducted before flowering, focuses on foliage and plant type traits to distinguish true-to-type plants. The second inspection, during the flowering and fruiting stage, examines a range of flower, fruit, and plant characteristics. The third inspection evaluates the color and shape of mature fruit, as well as the appearance of the fruit's shoulders, to confirm the authenticity of the plants.

Breeder seed production

1. Cultivate approximately 75-100 individual plant progenies, each comprising at least 20-25 plants, with breeder seed lines positioned every 25 progeny rows for comparison. Adjust the number of progenies as needed based on breeder seed requirements. Maintain a spacing of 100 cm × 50 cm between rows and plants.
2. Before flowering, identify and remove plant progenies that do not meet varietal purity standards based on plant type. During the immature and mature fruit stages, evaluate fruit size, shape, color, and overall performance. Eliminate plants exhibiting symptoms of early blight, leaf spot, and mosaic (TMV) to prevent seed-borne diseases.
3. Collect seeds separately from approximately 75 optimal plants to repeat the aforementioned cycle. The harvested seed from each plant is adequate to sustain the process for breeder seed production over the next 2-3 years, provided appropriate drying and storage procedures are followed.
4. Harvest true-to-type progenies in bulk to acquire breeder seed for subsequent seed multiplication.

Foundation Seed Production

1. Utilize the bulk seed collected from the selected progenies, as previously mentioned, for foundation seed production, maintaining the same spacing parameters employed for breeder seed production.
2. Identify and eliminate rogue plants displaying off-type foliage and other undesirable traits before flowering. Subsequent inspections should evaluate overall plant performance, including fruit shape, size, color, and internal characteristics, to ensure compliance with varietal descriptions.
3. Detect and remove diseased plants affected by early blight, leaf spot, and mosaic (TMV), ensuring that the percentage of diseased plants remains within the maximum limit specified by seed certification regulations.
4. Field standards: The minimum seed certification standards in India have been documented by Tunwar and Singh (1988) as follows.

DUS guidelines in tomato

The implementation of the Plant Varieties Protection and Farmers' Rights Act, commonly known as the PPV and FR Bill, in 2001 by the Government of India, aims to safeguard new varieties and germplasm. To qualify for protection under this Act, varieties must undergo evaluation through DUS (Distinctness, Uniformity, and Stability) and VCU (Value for Cultivation and Use) tests. Therefore, the differentiation of tomato varieties, particularly through examination of plant or seed morphology, is increasingly vital to uphold the rights of breeders and farmers and ensure the genetic purity of varieties, a crucial aspect of seed quality. Taxonomical descriptors for tomatoes are outlined by international bodies like the International Union for Protection of New Plant Varieties (UPOV, 1992). These descriptors hold traditional significance and serve as a classical taxonomic approach for variety identification (Table 4). Additionally, developing identification keys based on these morphological traits could establish a comprehensive database for cultivar identification and genetic purity testing.

Table 4: DUS descriptors in tomato

Sl. No	Descriptors
1	Seedling: Anthocyanin colouration of hypocotyl
2	Leaf: Intensity of green colour
3	Plant: Growth type
4	Stem: Pubescence
5	Stem: Anthocyanin colouration of upper third portion
6	Stem: Length of internode between 1st and 4th inflorescence (for indeterminate varieties) (cm)
7	Stem: Length of internode between 1st and 4th inflorescence (for determinate varieties) (cm)
8	Leaf: Length (cm)
9	Leaflet: Length (cm)
10	Leaf: Width (cm)
11	Leaflet: Width (cm)
12	Leaf: Serration
13	Leaf: Structure
14	Leaf: Attitude in relation to main stem (in middle third of plant)
15	Leaf: Attitude of petioles of leaflets in relation to main axis
16	Inflorescence: Type (2nd and 3rd truss)
17	Plant: Number of inflorescence on main stem (side shoots to be ignored) (for determinate varieties only)
18	Flower: Fasciations (1st flower of inflorescence)
19	Flower: Pubescence of style
20	Flower: Colour
21	RHS Colour Chart No.
22	Flower: Anther colour
23	RHS Colour Chart No.
24	Flower: Nature of stigma

25	Flower: Stigma
26	Flower: Calyx size (cm)
27	Peduncle: Abscission layer
28	Jointed peduncle: Length (from abscission layer to calyx) (cm)
29	Time of flowering (50% of the plants with at least one open flower from seed sowing) (days)
30	Fruit: Intensity of green colour (before maturity)
31	Fruit: Green shoulder (before maturity)
32	Fruit: Size (average weight of 10 fruits) (g)
33	Fruit: Length (cm)
34	Fruit: Width (cm)
35	Fruit: Shape in longitudinal section
36	Fruit: Ribbing at peduncle end
37	Fruit: Cross section
38	Fruit: Depression at peduncle end
39	Fruit: Size of scar around peduncle (diameter) (cm)
40	Fruit: Size of blossom scar
41	Fruit: Shape at blossom end
42	Fruit: Size of core in cross section(in relation to total diameter) (mm)
43	Fruit: Thickness of the pericarp (cm)
44	Fruit: Number of locules
45	Fruit: Colour at maturity
46	RHS Colour Chart No.
47	Fruit: Colour of flesh at maturity
48	RHS Colour Chart No.
49	Fruit: Firmness (kg/cm2)
50	Time of maturity (from seed sowing)
51	Fruit: Total soluble solids (°Brix)

Varieties

General Requirements

The seed fields must be kept separate from the contaminants listed in column 1 of the table, maintaining the specified distances as indicated in columns 2 and 3.

Characters	Isolation distances (meters)	
	Foundation (FS)	Certified (CS)
1	2	3
Plot with other breeding materials of Tomato	50	25
Fields of the same variety not meeting certification purity requirements.	50	25
Specific Requirements		
Maximum permitted (%)*		
	Foundation	Certified
Off types	0.10	0.20
Seed borne diseases/internally seed affecting factors **	0.10	0.50

- Maximum allowed at final inspection ** Seed-borne diseases: Leaf spot (Stemphylium solani Weber), Early blight (Alternaria solani Sorauer), Tobacco Mosaic Virus (TMV).

Seed standards of tomato

Futures	Class of seeds with specific features	
	Foundation	Certified
Pure seeds (maxi)	98.0%	98.0%
Inert matters (maxi)	2.0%	2.0%
Other crop seed (maxi)	5/kg	10/kg
Weed seeds (maxi)	None	None
Germination (mini)	70%	70%
Moisture (maxi)	8.0 %	8.0 %
For vapor-proof (Vp) containers (maxi)	6.0%	6.0%

Hybrids

General Requirements

Contaminants/Non acceptance	Min. distance (mtrs)	
	Foundation seeds (FS)	Certified seeds (CS)
Fields of other varieties of the same crop	200	100
Fields of the same hybrid (code designation) not meeting varietal purity requirements for certification.	200	100
Seed parent and pollen parents maintained in different blocks		5

Specific requirements

Specific needful features		
Features	**Min. permitted (%)***	
	Foundation seeds	**Certified Seeds**
Seed parent having off types in	0.010	0.050
Pollinizers having off types	0.010	0.050
** Fertile blocks in seed parent	0.050	0.10
*** Seed borne diseases affected plants	0.10	0.50

*Standards for off-types and fertile segregants in the seed parent must be met during and after flowering, and for seed-borne diseases at the final inspection. ** This condition is applicable when utilizing a male sterile line for hybrid seed production. ***Seed-borne diseases include early blight, leaf spot, and tobacco mosaic virus.

Seed Standards for Purity Maintenance

Traits	**Each classes of seeds with standards**	
	Foundation seeds	**Certified seeds**
1	2	3
Pure seed (min)	98.0%	98.0%
Inert matter (maxi)	2.0%	2.0%
Off types or other crop seeds (maxi)	5/kg	10/kg
Weed seeds (maxi)	None	None
Germination (minimum)	70%	70%
Moisture (maximum)	8.0%	8.0%
For vapor-proof (VP) containers (maxi)	6.0%	6.0%

Certified seed lots produced through emasculation and hand pollination will undergo a grow-out test and must meet the following minimum genetic purity standards:

Classes of seeds	Genetic purity (%) (minimum)*
Certified seeds	90.0%

*During the grow-out test, off-type plants (excluding selfed plants), including sergeants, outcrosses, and plants of other varieties, should not exceed 1.50% of the total 10.0% of plants designated for selfed plants.

Harvesting and seed extraction procedures

The tomato crops are harvested for seed when the majority of fruits reach ripeness but before significant rotting occurs, as the seed reaches physiological maturity at the red ripe stage. Seed obtained from unripe green, yellow, or sun-scalded fruits shows poor germination, and during seed harvest, diseased or rotten fruits should be discarded.

1. Tomato seeds are enveloped in a mucilaginous substance, complicating seed separation. Various methods exist to eliminate this mucilage.
2. Fermentation: Tomato fruits are pulped and left to ferment in wooden, earthen pots, or non-corrosive metal containers for 2-3 days, preferably during May-June in North India. Regular stirring prevents fungal growth.
3. Acid Treatment: Pulped tomato fruit is treated with commercial-grade hydrochloric acid at 10-15 ml per kg to dissolve the mucilage. Following stirring, the seeds are washed to eliminate pulp and skin residues. In instances of bacterial canker, the seeds are soaked in 0.8 percent acetic acid for 24 hours to eradicate the pathogen.
4. Juice Extraction: A juice extraction machine separates tomato seeds and pulp from the juice. The seeds are then washed to eliminate any remaining pulp and skin. If necessary for bacterial canker prevention, seeds are treated with acetic acid.
5. Mechanical Extraction: The process involves using an Axial Flow Vegetable Seed Extraction Machine for tomato seed extraction. Collected seed material is submerged in water to allow heavier seeds to settle at the bottom. Floating tomato pulp and skin fragments are removed, and the seeds are washed multiple times. Treatment with hydrochloric acid is applied, followed by further washing, ensuring removal of the seed's mucilaginous layer.

Drying and Storage

To facilitate rapid drying, tomato seeds undergo thorough washing before being spread thinly on drying floors or trays with screen bottoms, positioned outdoors to maximize sunlight exposure. In temperatures exceeding 40°C during summer, it is advisable to dry them under partial shade to prevent overheating and reduce the risk of seed germination reduction. Alternatively, under controlled conditions, the seeds undergo an initial drying phase at a low temperature (38°C) for several hours, followed by a gradual temperature rise to 40°C, ensuring that seed viability remains intact.

Properly dried tomato seeds should have no more than 8 percent moisture content. Packaging and storing the dried seed in a cool, well-ventilated, and dry environment is essential. If stored in moisture-proof or moisture-resistant containers, the moisture content of the seeds should not surpass 6 percent. Excessive moisture within sealed storage containers can rapidly diminish seed viability, especially in ambient or high-temperature conditions. With careful and proper storage, even under ambient conditions, the seed can retain its viability for 2 to 3 years, even when stored under ambient conditions.

References

Adhikari, T. B., Siddique, M. I., Louws, F. J., Sim, S. C., & Panthee, D. R. (2023). Molecular mapping of quantitative trait loci for resistance to early blight in tomatoes. Frontiers in Plant Science, 14, 1135884.

Al-Saikhan, M. S., Rezk, A. A., & Shalaby, T. A. (2020). Evaluation of TY-2 gene ability for resistance to three strains of tomato yellow leaf curl virus-like viruses in segregated populations of tomato. Fresenius Environmental Bulletin, 29(5), 3961-3969.

Aslam, M. N., Mukhtar, T., Hussain, M. A., & Raheel, M. (2017). Assessment of resistance to bacterial wilt incited by Ralstonia solanacearum in tomato germplasm. Journal of Plant Diseases and Protection, 124, 585-590.

Bai, Y., & Lindhout, P. (2007). Domestication and breeding of tomatoes: What have we gained and what can we gain in the future? Annals of Botany, 100(5), 1085-1094.

Ballester, A. R., Molthoff, J., de Vos, R., Hekkert, B. T. L., Orzaez, D., Fernández-Moreno, J. P., Tripodi, P., Grandillo, S., Martin, C., Heldens, J., & Ykema, M. (2010). Biochemical and molecular analysis of pink tomatoes: Deregulated expression of the gene encoding transcription factor SlMYB12 leads to pink tomato fruit color. Plant Physiology, 152(1), 71-84.

Chu, Y. H., Jang, J. C., Huang, Z., & van der Knaap, E. (2019). Tomato locule number and fruit size controlled by natural alleles of lc and fas. Plant Direct, 3(7), e00142.

Collins, E. J., Bowyer, C., Tsouza, A., & Chopra, M. (2022). Tomatoes: An extensive review of the associated health impacts of tomatoes and factors that can affect their cultivation. Biology, 11, 239.

El-Sappah, A. H., Islam, M. M., El-Awady, H. H., et al. (2019). Tomato natural resistance genes in controlling the root-knot nematode. Genes, 10, 92.

Fentik, D. A. (2017). Review on genetics and breeding of tomato (Lycopersicon esculentum Mill.). Advances in Crop Science and Technology, 5(5), 306.

Foolad, M. R., Zhang, L. P., & Lin, G. Y. (2001). Identification and validation of QTLs for salt tolerance during vegetative growth in tomato by selective genotyping. Genome, 44(3), 444-454.

Gill, U., Scott, J. W., Shekasteband, R., Ogundiwin, E., Schuit, C., Francis, D. M., Sim, S. C., Smith, H., & Hutton, S. F. (2019). Ty-6, a major Begomovirus resistance gene on chromosome 10, is effective against Tomato yellow leaf curl virus and Tomato mottle virus. Theoretical and Applied Genetics, 132, 1543-1554.

Gonda, I., Ashrafi, H., Lyon, D. A., Strickler, S. R., Hulse-Kemp, A. M., Ma, Q., Sun, H., Stoffel, K., Powell, A. F., Futrell, S., & Thannhauser, T. W. (2019). Sequencing-based bin map construction of a tomato mapping population, facilitating high-resolution quantitative trait loci detection. The Plant Genome, 12(1), 180010.

Haanstra, J. P. W., Wye, C., Verbakel, H., Meijer-Dekens, F., Van den Berg, P., Odinot, P., Van Heusden, A. W., Tanksley, S., Lindhout, P., & Peleman, J. (1999). An integrated high-density RFLP-AFLP map of tomato based on two Lycopersicon esculentum × L. pennellii F1 populations. Theoretical and Applied Genetics, 99, 254-271.

Jia, M. (2019). Genetic characterization and mapping of late blight resistance genes in the wild tomato accession PI 270443 (Doctoral dissertation, The Pennsylvania State University).

Knapp, S. (2002). Tobacco to tomatoes: A phylogenetic perspective on fruit diversity in the Solanaceae. Journal of Experimental Botany, 53(377), 2001-2022.

Li, W., Li, Y., Liang, Y., Ni, L., Huang, H., Wei, Y., Wang, M., Zhang, L., & Zhao, L. (2022). Generating novel tomato germplasm using the ancestral wild relative of Solanum pimpinellifolium. Horticulturae, 9(1), 34.

Liedl, B. E., Labate, J. A., Stommel, J. R., Slade, A., & Kole, C. (Eds.). (2013). Genetics, genomics, and breeding of tomato. CRC Press.

Liu, B., Song, L., Deng, X., Lu, Y., Lieberman-Lazarovich, M., Shabala, S., & Ouyang, B. (2023). Tomato heat tolerance: Progress and prospects. Scientia Horticulturae, 322, 112435.

Miller, E. C., Giovannucci, E., Erdman, J. W., Bahnson, R., Schwartz, S. J., & Clinton, S. K. (2002). Tomato products, lycopene, and prostate cancer risk. Urologic Clinics, 29(1), 83-93.

Ning, Y., Wei, K., Li, S., Zhang, L., Chen, Z., Lu, F., & Huang, Z. (2023). Fine mapping of fw6.3, a major-effect quantitative trait locus that controls fruit weight in tomato. Plants, 12(11), 2065.

Osei, M. K., Prempeh, R., Adjebeng-Danquah, J., Opoku, J. A., Danquah, A., Danquah, E., Blay, E., & Adu-Dapaah, H. (2018). Marker-assisted selection (MAS): A fast-track tool in tomato breeding. In Recent advances in tomato breeding and production (pp. 93-113).

Panthee, D. R., & Chen, F. (2010). Genomics of fungal disease resistance in tomato. Current Genomics, 11(1), 30-39.

Panthee, D. R., Brown, A. F., Yousef, G. G., Ibrahem, R., & Anderson, C. (2013). Novel molecular marker associated with Tm2a gene conferring resistance to tomato mosaic virus in tomato. Plant Breeding, 132(4), 413-416.

Peralta, I. E., Knapp, S., & Spooner, D. M. (2006). Nomenclature for wild and cultivated tomatoes. Tomato Genetics Cooperative Report, 56(1), 6-12.

Peralta, I. E., Spooner, D. M., & Knapp, S. (2008). Taxonomy of wild tomatoes and their relatives (Solanum sect. Lycopersicoides, sect. Juglandifolia, sect. Lycopersicon; Solanaceae).

Tanksley, S. D., & Rick, C. (1980). Isozymic gene linkage map of the tomato: Applications in genetics and breeding. Theoretical and Applied Genetics, 58, 161-170.

Tanksley, S. D., Ganal, M. W., Prince, J. P., De Vicente, M. C., Bonierbale, M. W., Broun, P., Fulton, T. M., Giovannoni, J. J., Grandillo, S., & Martin, G. (1992). High-density molecular linkage maps of the tomato and potato genomes. Genetics, 132(4), 1141-1160.

Tieman, D. M., Zeigler, M., Schmelz, E. A., Taylor, M. G., Bliss, P., Kirst, M., & Klee, H. J. (2006). Identification of loci affecting flavour volatile emissions in tomato fruits. Journal of Experimental Botany, 57(4), 887-896.

Tigist, M., Workneh, T. S., & Woldetsadik, K. (2013). Effects of variety on the quality of tomato stored under ambient conditions. Journal of Food Science and Technology, 50, 477-486.

Tiwari, J. K., Yerasu, S. R., Rai, N., Singh, D. P., Singh, A. K., Karkute, S. G., Singh, P. M., & Behera, T. K. (2022). Progress in marker-assisted selection to genomics-assisted breeding in tomato. Critical Reviews in Plant Sciences, 41(5), 321-350.

Valpuesta, V. (Ed.). (2002). Fruit and vegetable biotechnology. Woodhead Publishing.

Van der Knaap, E., Chakrabarti, M., Chu, Y. H., Clevenger, J. P., Illa-Berenguer, E., Huang, Z., Keyhaninejad, N., Mu, Q., Sun, L., Wang, Y., & Wu, S. (2014). What lies beyond the eye: The molecular mechanisms regulating tomato fruit weight and shape. Frontiers in Plant Science, 5, 227.

Vrebalov, J., Ruezinsky, D. M., Padmanabhan, V., White, R., Medrano, D., Drake, R., Schuch, W., & Giovannoni, J. (2002). A MADS-box gene necessary for fruit ripening at the tomato ripening-inhibitor (rin) locus. Science, 296, 343-346.

Zhang, D., Li, H., Liu, G., Xie, L., Feng, G., & Xu, X. (2023). Mapping of the Cladosporium fulvum resistance gene Cf-16, a major gene involved in leaf mold disease in tomato. Frontiers in Genetics, 14, 1219898.

2

Brinjal

***H.A. Bhargavi*[1]*, T.H. Singh*[1]*, N. Pradeepkumara*[2]*, and K.H. Sulochana*[1]**

[1]*Division of Vegetable Crops, ICAR-Indian Institute of Horticultural Research Bengaluru-560089, Karnataka*

[2]*ICAR- Central Institute of Temperate Horticulture, Regional Station, Dirang-790101 Arunachal Pradesh*

Abstract

Advances in seed production techniques have revolutionized brinjal cultivation, enhancing productivity, quality, and sustainability. Traditional methods, such as manual emasculation and pollination, have been supplemented and in some cases replaced by innovative approaches, contributing to increased efficiency and reduced labor costs. Cytoplasmic male sterility (CMS) and grafting technologies have emerged as valuable tools, offering novel avenues for enhancing breeding efficiency and managing soil-borne pathogens. These advancements have facilitated the development of high-yielding F1 hybrids and improved pure-line varieties with enhanced resilience to pests and diseases. Furthermore, the adoption of precision agriculture techniques including controlled environment agriculture and hydroponics, has enabled year-round seed production, mitigating the impact of seasonal fluctuations and adverse weather conditions. Integrated pest management (IPM) strategies have been integrated into seed production protocols, minimizing chemical inputs and promoting environmentally sustainable practices. Additionally innovations in seed extraction methods, such as mechanical extraction and dry extraction techniques, have streamlined the seed processing workflow resulting in higher seed purity and improved germination rates.

Introduction

Brinjal (*Solanum melongena* L.; 2n=24) is one of the essential vegetable crop grown in tropical and subtropical regions. Around 9,000 to 10,000 years ago the brinjal we know today was derived from its wild ancestor, *S. insanum* L. Ranked sixth globally in terms of production, brinjal follows tomato, onion watermelon, cucumber, and cabbage as one of the most significant vegetable crops. The scarlet eggplant (*S. aethiopicum* L.) and the gboma eggplants (*S. macrocarpon* L.), other two cultivated eggplant species which although less recognized, hold significant local significance within sub-Saharan Africa. Brinjal is a good source of vitamins and dietary minerals, fibres, proteins and anthocyanins, with a high antioxidant properties and low caloric value (Oladosu *et al.*, 2021). Regarding bioactive compounds, it contains high amount of chlorogenic acid and anthocyanins. Further it is associated with multiple health benefits. It is primarily cultivated for the fresh market with limited cultivation for processing purposes. Due to the economic and nutritional importance, breeding endeavors prioritize the development of high-yielding varieties, primarily focusing on F_1 hybrids with key traits in brinjal (Taher *et al.*, 2017). Hybrids have gained greater popularity compared to open-pollinated (OP) varieties due to their heightened fitness and vigor. They hold greater economic value compared to enhanced OP varieties.

The primary aims of brinjal breeding programs include improved fruit quality enhanced shelf life, boosting yield performance through heterosis breeding and incorporating pest and disease resistances from wild relatives. The diverse consumer preferences across different regions necessitate the development of numerous high-yielding F_1 hybrids for brinjal cultivation. There has been a steady increase in cultivated eggplant production, attributed to a 50% rise facilitated by the availability of high-yielding varieties and hybrids (Kumari *et al.*, 2020). Similarly, in recent years, there has been a significant surge in farmers' inclination towards eggplant hybrids. Researchers are also prioritizing the development of high-yielding eggplant hybrids to meet yield targets and satisfy market demands. The production of hybrid seeds is relatively straightforward and cost-effective in brinjal. Generally, brinjal hybrid seeds are generated through the technique of manual emasculation and pollination. The expense associated with hybrid seed production is relatively low compared to other vegetables, and this cost can be further mitigated through the adoption of male sterile lines.

Floral Biology

Eggplant flowers occur singly or in clusters of 2 to 5. They are complete hermaphroditic, and radially symmetrical. The calyx consists of five lobes fused petals, and remains attached persistently, forming a cup-shaped structure at the base. The corolla has five lobes, fused petals. Typically purple, although it can vary from greenish-white to white. There are five free stamens, situated at the corolla's throat, with cone-shaped anthers that open at the apex. The ovary is located below the floral parts, consisting of two fused carpels with basal placentation. Heterostyly, the variance in the positioning of the stigma relative to the anther tips, is a prevalent trait. Eggplant exhibits four flower types determined by the length of styles: (i) long-styled, featuring a large ovary, (ii) medium-styled, with a medium-sized ovary, (iii) pseudo short-styled, showcasing a rudimentary ovary, and (iv) true short-styled, displaying a very rudimentary ovary. In various cultivars, as well as within different flowers of certain cultivars, the positioning of stigmas relative to stamens can differ. Stigmas may be positioned above, at the same level as, or below the stamens. Long-styled flowers typically exhibit fruit setting rates ranging from 70 to 85%, while medium-styled flowers show approximately 55% fruit setting. Pseudo-short-styled and true short-styled flowers usually do not result in fruit setting. Eggplant primarily undergoes self-pollination, although cross-pollination rates have been observed to reach as high as 29%, thus categorizing it as occasionally cross-pollinated. Major pollinators include bumblebees (*Bombus* sp.) and honeybees (*Apis* sp.). Self-pollination is favoured in the absence of pollinators. Anthesis typically commences around 7:00 a.m. and continues until 11:00 a.m., with the peak period occurring between 8:30 and 10:30 a.m. Anthers generally release pollen approximately half an hour after flower opening. Receptivity spans from the day preceding flower opening in summer to 2-3 days in winter, under field conditions.

Breeding Objectives

Breeding programs for brinjal prioritize the development of varieties with enhanced fruit yields, early maturity, excellent quality, plant architecture and resilience against diseases. Essential quality characteristics include fruit color, seed-to-pulp ratio, reduced cooking duration, and minimal solanine content. The main aim of traditional breeding is to tackle issues such as insect pests, diseases, and environmental stresses including drought, heat, salinity, and cold.

Although manual emasculation and pollination are effective methods, they are labour-intensive, leading to the exploration of cytoplasmic male sterility to enhance breeding efficiency. Additionally, grafting technology has emerged as a valuable tool, particularly in regions affected by soil-borne pathogens. Successful grafting has been demonstrated between *Solanum melongena* (brinjal) and various rootstocks, offering a promising solution for managing these challenges. Pure-line selection, pedigree breeding, and backcrossing are commonly utilized methods, while F_1 hybrids are increasingly favoured for their vigour, uniformity, and increased yield. The emergence of heterosis in brinjal has led to the development of initiatives to create hybrids with enhanced productivity with the use of selected inbred lines. These hybrids not only offer improved yields and quality, but also contribute significantly to the escalating demand for brinjals both locally and internationally. The selection of brinjal varieties holds significant importance in cultivation due to varying local preferences across regions and even districts. A variety prized for its specific fruit color and size, commanding a premium price in one market, could be completely disregarded in another locale. As farmers strive to maximize returns from their produce, careful consideration of market demands is paramount in selecting the most suitable variety for cultivation in this crop.

Varietal Achievements

Table 1. Improved cultivars of brinjal developed by public sector in India

Cultivar	Special attributes	Source
CO.1	The oblong fruits of this variety exhibit a pale green hue against a white background. Even upon reaching full maturity, the fruits remain soft-seeded.	TNAU, Coimbatore
CO.2	The fruits display a slight oblong shape and feature dark purple streaks against a pale green background, with no spines present on the calyx surface.	TNAU, Coimbatore
MDU. 1	Large sized fruit with less seed content	TNAU, Coimbatore
PKM . 1	The variety derived through induced mutation from a local variety known as Puzhuthikathiri. Small fruit with green stripes.	TNAU, Coimbatore
PPI (B)	The fruits are elongated, pale green in colour, with fewer seeds and a less bitter taste.	TNAU, Coimbatore
TNAU Brinjal VRM 1	The fruits exhibit an oval shape, boasting a glossy pink coloration with a greenish tinge at the distal end. They demonstrate resilience against leaf spot, verticillium wilt, and epilachna beetle. This variety is characterized by its cluster-bearing nature.	TNAU, Coimbatore
KKM 1	The plants are compact and produce small-sized, egg-shaped fruits in clusters of 2-4, each adorned with a green calyx.	TNAU, Coimbatore
Pusa Purple Long	The fruits are smooth, glossy, and light purple in colour, measuring 25-30 cm in length.	IARI, New Delhi
Pusa Purple Round	It is a round-shaped variety of eggplant. The plant features a tall stature, characterized by a thick stem displaying hues of greenish-purple and highly lobed, green-colored leaves.	IARI, New Delhi
Arka Neelanchal Shyama	The fruits exhibit a round shape and are green with a hint of light purple. They possess a moderate tolerance to Phomopsis blight.	IIHR, Bengaluru
Arka Unnathi	This variety is known for its high yield and resistance to bacterial wilt. The fruits are elongated and medium green, with a fleshy green calyx.	IIHR, Bengaluru

Arka Avinash	This particular variety boasts high yields and resistance to bacterial wilt. Its fruits are elongated and green, accompanied by a fleshy green calyx.	IIHR, Bengaluru
Arka Neelkanth	The fruits are tender with a slow seed maturation process and lack bitter principles. Additionally, they exhibit resistance to bacterial wilt.	IIHR, Bengaluru
PB-67	It is Early maturing bacterial wilt and Phomopsis resistant variety	GBAU&T, Pantnagar
PB-70	It is Phomopsis blight, bacterial wilt and shoot and fruit borer resistant variety	GBAU&T, Pantnagar
Pant Samrat	Phomopsis blight resistant variety	GBAU&T, Pantnagar
F_1 Hybrids		
Arka Anand	An F_1 hybrid resistant to bacterial wilt was created by crossing IIHR-3 with IIHR-322.	IIHR, Bengaluru
Arka Navneeth	Purple colour fruit with free from bitterness	IIHR, Bengaluru
Pusa Anmol	A high yielding purple coloured hybrid	IARI, New Delhi
Pusa Hybrid - 5	Glossy purple fruit with good yield	IARI, New Delhi
Pusa Hybrid - 6	Purple fruit with early maturity	IARI, New Delhi
COBH.1	Dark violet coloured fruit with high ascorbic acid content	TNAU, Coimbatore
COBH. 2	Glossy violet colored fruit with moderate tolerance to shoot and fruit borer	TNAU, Coimbatore
Azad Hybrid	Purple coloured fruit with high shelf life	C.S.Azad Agricultural University, Kanpur
Hisar Shyamal (H8)	Purple fruit with bacterial wilt resistance	HAU, Hisar

Table 2. Brinjal cultivars developed by private sector and being cultivated in India

Company	**Leading hybrids**
Mahyco, Pvt Ltd	MAHY 80 (MHB 80), MAHY 205, MHBJ 110, MAHY 343, MHB – 1, MHB 9, MHB – 20, MAHY 99, MAHY green,
Bejo Sheetal Seeds	Rasika
Seminis Seeds	Shamli
VNR Seeds	VNR-51C, VNR Navina, VNR Simran
Hindustan Seeds Private Limited	Brinjal Sungrow 132 Gold, Brinjal Navkiran
Farmson Biotech	Fb-Ragini
Hi-tech genetic seeds	Sunidhi 5552
Iris Hybrid Seeds	Eklavaya

Heterosis Breeding in Brinjal

Heterosis presents opportunities for enhancing productivity, early maturity, uniformity, quality, adaptability, and the rapid incorporation of dominant genes for disease and pest resistance. Leveraging heterosis in brinjal cultivation can lead to the development of superior hybrid combinations, meeting the increasing demand for brinjal driven by its bioactive properties. The earliest documented instance of heterosis in eggplant was reported by Munson in 1892. Subsequently, Halsted in 1901 noted that in one cross, the fruit size doubled compared to the parental line, along with increased yield. Harnessing hybrid vigor has emerged as a promising approach for enhancing brinjal production. Pal and Singh (1949) documented a substantial increase in yields ranging from 48.8% to 56.6% in brinjal hybrids compared to the superior parent. Kakizaki (1931) was among the first to recognize the commercial potential of hybrids, highlighting their superior yields compared to standard varieties during the years 1923-1926. The utilization of heterosis or hybrid vigor has been a significant strategy for enhancing eggplant cultivation since its inception. The list of popular public and private sector hybrids are mentioned in table 3.

Table 3: Popular hybrids developed by public and private sector organisations

Crop	Public Bred Hybrids	Private Sector Hybrids
Brinjal	Arka Anand Arka Navneet HABH-8 PB-70 DBL-02 DBHL-20 PHBL-51 PBHSR-31 HABR-21 PBHL-52 COBH.1 COBH. 2	Rasika Shamli VNR-51C VNR-218 MHB 10 Azad Hybrid MHB 39

Seed Production Technique in Brinjal

Eggplant, a vital vegetable crop widely cultivated in tropical and subtropical regions, holds significant importance in fresh markets and to a lesser extent, in processing. Currently, there is a growing emphasis on enhancing eggplant breeding and production. To maximize production, implementing improved cultural practices alongside high-quality seed selection is essential. The process of producing high-quality eggplant seeds involves the following steps.

Emasculation and Pollination

Brinjal is a self-pollinated crop with 5-10 % cross pollination. Flower is hermaphrodite with 5-6 petals coalesced. In brinjal, self-pollination is typically more prevalent than cross-pollination. Although it is categorized as an often-cross-pollinated crop, the reported average cross-pollination rate ranges from 6.75% to 48%. Bees are often chief insect-pollinators of brinjal. To prepare for emasculation, a healthy, well-developed bud from the central part of the plant, preferably with a long or medium style, is carefully chosen. This bud is delicately opened using fine-pointed forceps one or two days before it naturally opens, ensuring the removal of all five anthers. Flower buds anticipated to bloom the following day from the male parent are gathered in the evening prior and stored in polythene covers. The next morning, the pollen is separated from the anthers and collected in a container.

For pollination, freshly dehisced anthers are selected and vertically slit with a fine needle to ensure an adequate amount of pollen is available at the tip. The collected pollen from the male line is then transferred onto the stigmatic surface of the emasculated flower of the female line. Alternatively, the stigmatic surface can be dipped directly into the container holding the pollen. The process is meticulously labelled and covered with a small pollination bag (Figure 1). Bud pollination is ideally conducted between 7:30 and 10:30 a.m.

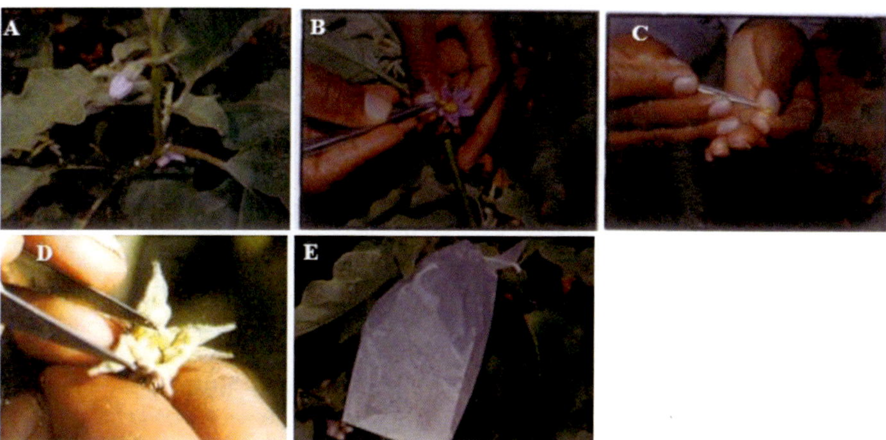

Figure 1: Emasculation and pollination in brinjal: **(A)** Unopened flower bud; **(B)** Removal of petals and anthers; **(C)** Removal of sepals; **(D)** Pollinate the flower after removing its stamen; **(E)** Bagging and tagging of pollinated flower.

Seed Production

To raise seed crop one should always use breeder or foundation seed. It should be obtained from the reliable source. The male and female parents are cultivated 200 meters apart to ensure isolation from other eggplant cultivars. Hybrid seed production involves manual emasculation and hand pollination of individual flowers. This process is facilitated by the presence of large flowers and is cost-effective because each fruit yields a reasonable quantity of seed. It is particularly advantageous when only a limited quantity of seed is needed for planting. For raising one hectare 250-300 seeds of pure line and 120g of hybrid seed will be sufficient the healthy nursery should be raised properly and carefully, preferably near a water source and either in poly house or in net house. The timely plant protection measures should be followed. Transplant only healthy seedling when they are 4-7 weeks old. Hardening of seedlings may be done by withholding the irrigation for avoiding the transplanting shocks and better establishment. Depending upon the fertility status of soil, season and nature of the variety, different spacing may be followed. Generally, a spacing of 90 cm × 50 cm followed for open pollinated varieties and 1.00m × 60cm spacing for hybrids varieties. Immediately after transplanting field may be irrigated. In places spacing between the rows is adjusted to meet the intercultural operation by mechanical means.

Land requirement: To ensure quality and quantity of produce, it's essential to choose a field free from volunteer plants for brinjal seed production. The selected field should not have cultivated solanaceous crops continuously in the past 3-4 years. It's preferable to select upland fields for brinjal cultivation to prevent waterlogging during the rainy season.

Soil and climate: An optimal soil for crop production is characterized by being well-drained, structurally stable, deep, and fertile, ranging from sandy to heavy clay with high pore volume. Ideally, the soil pH should fall within the range of 5.5 to 7.0 for optimal crop growth. Brinjal thrives in warm climates and requires a prolonged period of warmth for optimal growth, making it highly vulnerable to frost. The most favorable temperature range for successful brinjal production is a daily mean temperature of 13-21°C. Frost poses a significant threat to brinjal, with growth severely impacted when temperatures drop below 17°C. This crop can be cultivated successfully during both the rainy and summer seasons and can even be grown at elevations of up to 1200 meters above sea level.

Fertilizers: Fertilizer application involves distributing 60 kg of nitrogen along with the full amounts of phosphorus (100kg P_2O_5) and potassium (50kg K_2O) fertilizers at the basal stage. Subsequently, the remaining 60 kg of nitrogen is applied as a top dressing 30 days after transplanting. Finally, a third dose of 60 kg of nitrogen is administered 50 days after transplanting. Furrows and ridges across the slope are prepared. Furrows at appropriate spacing are opened and fertilizer mixture is applied in bands and earthed up. After lightly irrigating the field, holes are created at suitable intervals, and seedlings are planted along the sides or ridges, with one seedling per hill. Irrigation follows promptly after planting. Irrigation is given at 5 day intervals depending on the season & soil type. On 25th/30th day, after top dressing fertilizer on the sides or ridges, earthing up is to be done to bring plants to the middle of the ridge. Timely removal of harmful weeds and pulverizing the soil should be done for proper growth of the plants. However, use of weedicides can be helpful in reducing cost of cultivation and economize intercultural operations as labour input is very costly. The plant protection needs to be carried out timely following appropriate method.

Temperature and fertilizer requirements

Crop	Optimum temperature requirements			Fertilizer requirement(kg/ha)			Soil PH
	Seed germination	Plant growth	Fruit set	N	P_2O_5	K_2O	
Brinjal	20-32 °C	13-21 °C	21-29 °C	150	100	50	6.5 – 7.5
Hybrid varieties	20-32 °C	13-21 °C	21-29 °C	200	100	100	6.5 – 7.5

Weed management: For weed management in brinjal crops, soil solarization involves the use of polyethylene film with a thickness of 50-100 microns, which has proven highly effective. If available, crop residues such as rice, sorghum, wheat, maize can be utilized, or otherwise disposed of as waste. Initially, hand weeding can be substituted with chemical weeding, which can later be complemented with additional chemical weed control methods.

Isolation: The separation maintained between two seed crops or varieties of the same crop to safeguard genetic purity is termed as isolation distance. This distance is determined solely by factors such as pollination behavior, seed crop stages, hybrid varieties, pollinating agents, and pollen dispersal capability. Since there is considerable cross-pollination in brinjal, an appropriate distance is required between different varieties or from the same variety. A minimum isolation distance of 200 meters must be maintained to avoid genetic contamination.

Seed Certification Standards

The General Seed Certification Standards serve as the foundation, and in conjunction with the subsequent specific standards, they form the criteria for certifying brinjal seed.

Field Standards

(A) General requirements

Contaminants	Minimum distance (meters)	
	Foundation	Certified
Fields of other varieties	200	100
Fields of the same variety not conforming to varietal purity requirements for certification	200	100

(B) Specific requirements

Factor	Maximum permitted (%)*	
	Foundation	Certified
Off types	0.10	0.20
** Plants affected by seed borne diseases like Phomopsis blight (*Phomopsis vexans* (Sacc. & Syd.) Harter.)	0.10	0.50
Plants affected by little leaf but not seed borne diseases	0.50	2.0

Field Inspection and Rouging

For the assurance of seed purity, a minimum of three field inspections are recommended to detect and eliminate off-type plants. The initial inspection occurs pre-flowering, during which off-type plants are removed, effectively preventing the occurrence of outcrossing. Plants that deviate from the expected varietal characteristics in terms of plant type and leaf traits should be removed. Additionally, any plants exhibiting exceptionally early flowering should be uprooted. During the second inspection, conducted at the flowering stage, individual plants are examined for flower color along with other traits such as leaf and stem color, presence or absence of spines/thorns on leaves, as well as plant height and spread. Plants that do not match the varietal description should be eliminated. Any plants showing early or delayed flowering and fruit setting compared to the check rows should also be removed. The third inspection occurs when the fruits reach the edible stage. During this phase, individual plants' fruits are examined for size, color, and shape, alongside assessing the morphological traits of the plant. Even at this stage, removing off-type plants aids in preventing physical contamination in the seed.

Major Pest and Diseases and Their Control

Major Pests and Diseases	Control Measures
Shoot and fruit borer	1. Remove infested shoots by cutting them half an inch below the bore point and bury them deep in the soil. 2. Apply Malathion at a rate of 2ml per liter of water by spraying the crop.

Red spider mite	Apply Kelthane or Sulfur at a concentration of 2 ml or gm per Liter of water, with sprays administered at 7-day intervals.
Leaf eating beetle	1. If infestation is limited, manually pick larvae by hand. 2. Apply a dusting of a combination of ash and Sevin 50 WP onto plants. 3. Spray the crop alternatively with Sevin 50 WP at a rate of 3g and Endosulphon at a concentration of 2 ml per liter of water, with treatments administered at 7-day intervals.
Damping off	1. To achieve efficient control, apply thiram, Captaf, or Agrosan G.N. at a rate of 2 g per kg of seed for seed treatment. 2. Drench the nursery beds with a solution of Captaf at a concentration of 2 g per Liter of water.
Phomopsis blight	Spray the crop with Dithane Z-78 at a concentration of 2 g per Liter of water at intervals of 7 to 10 days.
Little leaf	1. Remove and incinerate diseased plants. 2. Apply root treatment using 100 PPM of Tetracycline, followed by three sprays at weekly intervals for 4-5 weeks after transplanting. 3. Spray with Metasystox or Ekalux at a concentration of 2 ml per Liter of water until fruit setting.
Fusarium wilt and Verticillium wilt	1. Apply a soil drench of Captan at a concentration of 0.3% to field plants. 2. Practice sanitation by removing crop debris, deep ploughing, and implementing appropriate crop rotation.

Harvesting of fruits: Brinjal fruits are left to ripen beyond the point of being edible before they are harvested for seed production. The timing of fruit harvest in Brinjal is crucial; seeds obtained from fruits harvested when they have turned completely yellow yield the highest fruit and seed yield per hectare, seed recovery percentage, germination rate, and vigor index. Harvesting the fruit earlier or later than this optimal stage has been observed to decrease both yield and seed quality.

Seed extraction: Two primary methods are employed for extracting eggplant seeds: wet extraction and dry extraction. Generally, wet extraction is preferred, particularly in large-scale seed production operations.

Wet extraction of seed: For initial eggplant seed extraction, the fruits are sliced, crushed, and the seeds separated from the remaining fruit pulp and debris. However, because eggplant fruit pulp tends to be dry, additional water is needed during and after crushing to enhance separation. Typically, the fruits are crushed, and the seeds are extracted through washing and sieving. It's advisable to commence extraction in the morning to ensure that seeds are at least partially dried by evening to prevent germination during the process. Usually, seeds are manually scooped out from the sliced or crushed fruits, which is a labor-intensive and unhygienic method. Alternatively, an axial-flow vegetable extracting machine (Verma and Singh, 1988) can be utilized for seed extraction from eggplant fruits. Mechanical extraction offers advantages of affordability, speed, and maintains seed quality without adverse effects.

Dry extraction of seed: In certain countries, over-ripe eggplant fruits are sun-dried until they shrink. In purple and purple-black cultivars, this drying process causes the skin color to fade to a coppery brown hue. Subsequently, the fruits are manually beaten, and the seeds are hand-extracted. Although this method is time-consuming and labor-intensive, it is employed in some regions for producing relatively small seed lots. Ripe fruits are collected over several weeks, and manual labor is utilized for the final extraction process. Following thorough washing, the seeds are promptly dried either on screen-bottom trays exposed to sunlight and air or through artificial drying methods. Alternatively, the seeds can be spread in thin layers on a threshing floor or tarpaulin for drying. The dried seeds, with minimal moisture content, should undergo proper processing, grading, packaging, and then stored in a cool and dry environment.

Seed storage: Seeds are dried to a safe moisture content and treated with either Captan or Thiram at a rate of 2 g/kg, can be stored for 15 months in containers that allow moisture vapor to pass through. Alternatively, when stored in containers impermeable to moisture vapor, they can be preserved for 30 months.

Seed yield: Seed yield is influenced by various factors such as variety, environmental conditions, cultural practices, crop season, and management techniques. On average, seed yield ranges around 150 kg per hectare, although high-yielding cultivars cultivated in regions with excellent crop management can achieve yields of up to 200 kg/ha. The 1000-seed weight typically averages around 5 g, although smaller-seeded cultivars may have a 1000-seed weight of 4 g.

Seed Standards

Factors	Minimum percentage allowed	
	Foundation	Certified
Pure Seed (minimum)	98.0%	97.0%
Inert matter (maximum)	2.0%	2.0%
Other crops seeds (maximum)	None	None
Weed seeds (maximum)	None	None

Germination (minimum)	70%	70%
Moisture (maximum)	8.0%	8.0%
For vapour-proof containers (maximum)	6.0%	6.0%

Distinctiveness, Uniformity and Stability (DUS) Guidelines in Brinjal

Brinjal is renowned for its wide array of vegetative, floral, and fruit characteristics, encompassing variations in shape, size, and color, as well as its versatility in culinary applications, whether consumed fresh, dried, or pickled. The coloration of its leaves, flowers, and fruits is influenced by the presence of different pigmentations. The green hue of brinjal's leaves and fruits is attributed to chlorophyll pigments, while the purple tint observed in seedlings, stems, leaves, flowers, and fruits is linked to the presence of anthocyanin pigments. The initial phase of maintenance breeding involves scrutinizing and categorizing morphological traits, laying the foundation for delineating the uniqueness and reliability of cultivars (PPV&FRA, 2009). Ensuring a stable genotype is crucial for identifying cultivars suitable for cultivation across varied environments. List of the DUS descriptors are described in table 4.

Table 4. DUS descriptors in brinjal

SI.No	Descriptors
1	Seedling: Anthocyanin colouration of hypocotyl
2	Seedling: Intensity of anthocyanin colouration of hypocotyl
3	Stem: Anthocyanin colouration
4	Stem: Intensity of anthocyanin colouration
5	Stem:Pubescence
6	Leaf: Length (cm)
7	Leaf: Width (cm)
8	Leaf: Margin
9	Leaf: Blistering
10	Leaf: Spininess
11	Leaf: Intensity of spininess
12	Leaf: Blade colour
13	Leaf: Intensity of colour of blade
14	Leaf: Colour of vein
15	Leaf: Intensity of colour of veins
16	Inflorescence: Number of flowers
17	Flower: Size
18	Flower: Colour
19	Flowering :Time(days after seed sowing)
20	Fruit: Length (cm)

SI.No	Descriptors
21	Fruit: Diameter (cm)
22	Fruit: Length / diameter ratio
23	Fruit: General shape
24	Fruit: Diameter of pistil scar (cm)
25	Fruit: Shape of apex
26	Fruit: Curvature (only for cylindrical types)
27	Fruit: Colour of skin at commercial harvesting
28	Fruit: Intensity of purple colour of skin
29	Fruit: Intensity of green colour of skin
30	Fruit: Stripes
31	Fruit: Density of stripes
32	Fruit: Patches
33	Fruit: Glossiness at harvest maturity
34	Fruit: Size of calyx
35	Fruit: Colour of calyx
36	Fruit: Intensity of colour of calyx
37	Fruit: Spininess of calyx
38	Fruit: Ribs
39	Fruit: Creasing of calyx
40	Fruit: Colour of flesh
41	Fruit: Length of peduncle (cm)
42	Fruiting: Pattern
43	Plant: Growth habit
44	Plant: Height (cm)
45	Plant: Spread (distance between two extremes leaf tips at widest point (cm)
46	Fruit: Colour of skin at physiological maturity
47	Time of physiological ripeness (days after fruit set)

References

Halsted BD. Experiments in crossing eggplants. New Jersey Agricultural Experiment hybrid vigour in the brinjal and bitter gourd. The Indian J Gene. Pl. Bree. 1901;6:19-33

Kakizaki Y. Hybrid vigour in eggplant and its practical utilization. Journal of Here. 1931;21:253-258.

Kumari, N., Kant, K., Akhtar, S., Sangam, S., Kumar, S., Singh, V.K., Kumar, B. and Kumar, R., 2020. Economic Feasibility of Varied Nitrogen and Potassium Application in Eggplant in Middle Gangetic Plains of Bihar. Int. J. Curr. Microbiol. App. Sci, 9(10), pp.1224-1229.

Munson WM. Notes on eggplants, Maine Agriculture Experimental Station. Annual Rep; c1892. p. 76-89.

Oladosu, Y., Rafii, M.Y., Arolu, F., Chukwu, S.C., Salisu, M.A., Olaniyan, B.A., Fagbohun I.K. and Muftaudeen, T.K., 2021. Genetic diversity and utilization of cultivated eggplant germplasm in varietal improvement. Plants, 10(8), p.1714.

Pal BP, Singh HB. Studies in hybrid vigour II. Notes on the Manifestation of hybrid vigour in brinjal and bittergourd. Indian J of Gen. and Pl. Bree. 1946;6:19-33.

PPV&FRA (2009). Test guidelines applied for all varieties, hybrids and parental lines of brinjal/ eggplant (Solanum melongena L.). Plant Variety J,03 (11), 185-196.

Taher, D., Solberg, S.Ø., Prohens, J., Chou, Y.Y., Rakha, M. and Wu, T.H., 2017. World vegetable center eggplant collection: origin, composition, seed dissemination and utilization in breeding. Frontiers in Plant Science, 8, p.1484.

3

Chilli and Capsicum

Ponnam Naresh[1], Smaranika Mishra[1], Sai Timma Rao[1] Satyaprakash Barik[2] and K. Madhavi Reddy[1]

[1]Division of Vegetable Crops, ICAR-IIHR, Bengaluru-560089, Karnataka

[2]CV Raman Global University, Bhubaneswar-752054, Odisha

Abstract

Chilli (Capsicum annuum L.) and bell pepper (Capsicum annuum var. grossum) are economically significant crops in India, known for their distinct attributes and culinary uses. Despite their shared progenitor, these crops exhibit varietal differences and require specific breeding strategies for hybrid seed production due to small flower size. The development of male sterile systems has revolutionized hybrid seed production, with the exploitation of Genetic Male Sterility (GMS) and Cytoplasmic Male Sterility (CMS) in chilli and bell pepper breeding. The market for chilli seeds in India is substantial, driven by the demand for high-value hybrid seeds. This chapter comprehensively reviews the botanical characteristics, genetic resources, and breeding objectives in Capsicum species, emphasizing traits like disease resistance, high yield, and stress tolerance. It also highlights varietal achievements in public and private sectors, showcasing improved cultivars and hybrids with specific attributes. The integration of molecular markers in male sterile genes facilitates efficient hybrid seed production, ensuring genetic purity and quality. Additionally, the package of practices for seed production in chilli underscores the importance of isolation, field standards, inspection, harvesting, seed processing, and quality standards to maintain seed purity and viability.

Introduction

Chilli and bell pepper (*Capsicum annuum* L.; 2n=24) constitute one of the most valuable commercial spice-cum-vegetable crop group grown in India and belongs to the family Solanaceae. Though chilli and bell pepper are considered to have same progenitor, they differ at variety level. Chilli is *Capsicum annuum* L. *annuum* whereas, bell pepper is *Capsicum annuum* var. *grossum*. Total of thirty five species have been reported in *Capsicum* (Carrizo García *et al*. 2016) among which five species namely *C. annuum*, *C. frutescens, C. baccatum C. chinense* and *C. pubescens* are cultivated species. The wild progenitor of cultivated chilli is *C. annuum* var. *glabriusculum* (Aguilar-Melendez *et al.*, 2009). Pungency in chilli is due to capsaicin which is a lipophilic chemical that produces a burning sensation when comes in contact with saliva. Chilli has earned a special position in Indian cuisine for its distinct spiciness and aroma. Bell pepper though not spicy but its attractive colour and aroma fetches good price in market. Because of its specific climatic requirement, it is grown mostly under protected condition and considered as a high value and low volume crop. Seed production in these crops are generally done through manual emasculation and pollination hence, development of hybrid seeds is tedious owing to small flower size and thereby increases the seed cost as well. However, with the development and use of male sterile systems in hybrid seed production of chilli, area under hybrids has been increased extensively. Both GMS and CGMS systems have been commercially exploited in hybrid development of chilli. Whereas, CGMS system is not stable in bell pepper and use of GMS for hybrid seed production is gaining momentum in bell pepper.

Chilli and capsicum market segments can be broadly grouped into four i.e. fresh market segment, fresh processing market segment, dried spice segment and industrial extracts (Reddy *et al.*, 2013). India has a significant share of 4% in the global seed market. The Indian market for vegetable seed is projected to grow at a CAGR of 7-8% for the forecast period between 2020-2025.The share of the public sector in seed production in the country reduced from 42.72 per cent in 2017-18 to 35.54 per cent in 2020-21, while the share of the private sector grew from 57.28 per cent to 64.46 per cent during the same period, highlighting the rising role of private companies in India's seed sector. *(Courtesy The Hindu: Business Line).* In India, currently, there are approximately 850 seed companies (small to large-sized) operating in India. Owing to high cost and low volume of hybrids seeds requirement, there is huge demand for chilli seed market in India. The current chilli hybrid seed market is approximately~50 tons per year, with an estimated value worth of 16 million US $ (Reddy *et al.*, 2014)

Botany and Domesticated Species

Capsicum annuum L., a highly diverse species, is an annual herb or shrub characterized by its tall and erect stature, typically reaching heights of 1-1.5 meters. It boasts a robust taproot system accompanied by numerous lateral roots. Its leaves, are arranged alternately and exhibit an ovate shape with dimensions ranging from 10-16 cm in length and 5-8 cm in width. These leaves feature acuminate apices and vary in colour from light to dark green.

The plant produces solitary flowers with cup-shaped calyces typically adorned with five conspicuous teeth. The corolla, which can range from white to purple, measures 8-15 mm in diameter and is characterized by 5-7 lobes. Its ovary is typically 2-4 locular, with a filiform style and capitate stigma. The fruit, known as a berry, varies significantly in shape, size, colour, and pungency level. Seeds are typically pale yellow, orbicular, and flattened.

Capsicum frutescens L., formerly considered an independent species, is now classified as a variety of *Capsicum annuum*. This perennial shrub features ovate-lanceolate leaves with gradually acuminate apices and obliquely rounded to acute bases. Flowers are borne in groups of 2-3 per node, with triangular, acute, cupular calyces and white-colored corollas with 5 lobes. Stamens are exerted, and the fruit gradually tapers towards the apex, varying in colour from green to white.

Capsicum baccatum L. is a perennial herb or shrub known for its erect growth habit and branching structure, reaching heights of up to 4.5 meters. Its branches are slightly striate and densely pubescent. The plant produces solitary leaves with short petioles and double whorled solitary flowers. The calyx is cupular, five-ribbed, and five-dentate, while the corolla is greenish-white to straw-white, with a yellow throat. The fruit, erect and glossy, has a rounded distal end.

Capsicum chinense L. is a perennial sub-shrub characterized by its woody base and stout taproot. The plant exhibits an erect and upright growth habit, with two to three primary branches and 4-6 secondary branches. Its leaves are unique, featuring a crinkled texture and an ovate shape with acute apices. Flowers are white with a tinge of light brown and are borne in clusters of 3-6. Anthers are blue or dark green, with purple filaments, and the ovary is perigynous, forming a hollow cup shape.

Capsicum pubescens L. is a perennial shrub that can reach heights of up to 4 meters and has a lifespan of up to 15 years, imparting a tree-like appearance. Its leaves have petioles ranging from 5–12 mm in length and ovate leaf blades measuring 5–12 cm in length and 2.5 to 4 cm in width. Flowers are solitary or in pairs, with blue-violet-coloured petals and dark, rugose seeds.

Crossability and Hybridization

In consideration of both agronomic and taxonomic interests, numerous intraspecific and interspecific crossings have been conducted by various researchers in the past. Research studies have demonstrated that fertile hybrids can be successfully obtained through crosses within the *Capsicum annuum* complex to varying degrees. However, attempts to produce fertile hybrids through crosses with *Capsicum pubescens* have been unsuccessful. Limited success has been observed in obtaining fertile hybrids through crosses between wild and semi-domesticated *C. annuum* var. *glabriusculum*, as well as domesticated *C. annuum* var. annuum (Hernández-Verdugo *et al.*, 2001).

Genetic Resources, Land Races and Improved Varieties

National Bureau of Plant Genetic Resources (NBPGR), New Delhi is the main center for germplasm exploration, collection and conservation which maintains 5180 accessions (http://www.nbpgr.ernet.in/PGR_Databases.aspx). National Institutes like ICAR-IIHR, Bengaluru has 2000 accessions (IIHR Annual Report, 2020). ICAR-IARI, New Delhi, ICAR-IIVR, Varanasi, Horticultural Research Station, Lam farm, Dr YSR Horticultural University, Guntur, Punjab Agricultural University, Ludhiana, and Horticultural Research Station, Devihossur, University of Horticultural Sciences and AICRP (Vegetable Crops) are actively involved in chilli germplasm collection, conservation and maintenance.

Floral Biology and Pollination Mechanisms

The flower of the Capsicum genus is hermaphrodite, hypogynous, and pentamerous. Optimal conditions for maximum pollen viability and fruit set range between 16-32°C. Relative humidity is also crucial, as low humidity (< 50% RH) often leads to flower and fruit drops. Dry heat poses a greater threat than wet heat in this crop. Typically, fruit reaches the mature green stage 35-50 days post-pollination. Capsicum flowers comprise calyx, corolla, androecium, and gynoecium. Most *Capsicum* species are self-compatible. The stigma's position varies; it may be slightly below (bell pepper), level with the anthers, or exerted beyond (chilli), with the latter increasing the likelihood of cross-pollination compared to bell pepper.

Given the diminutive size of chili flowers, it is essential to exercise caution during the breeding program and seed production to prevent uncontrolled pollination. For hybridization, unopened flower buds are chosen, and their petals are meticulously removed using alcohol-sterilized forceps to expose the reproductive organs. Following the emasculation of anthers, the stigma is carefully examined for potential contamination. Pollination is conducted by transferring pollen with a small paintbrush. Subsequently, the cross is labelled, and seeds are harvested post-fruit ripening.

Breeding objectives

The climate change scenario, has led to change in biotic and abiotic stresses patterns effecting the crop production, there by the objectives and traits to be focused. According to survey carried out by Schleiermachers and Lin

(2023) to prioritize the traits of importance for breeding and listed twenty-four traits as major traits to be focused in breeding chilli in the present scenario. Begomovirus (leaf curl virus), thrips resistance, high yield, hight temperature stress tolerance and anthracnose fruit rot resistance are top five major traits to be focused up on developing varieties and hybrids. Stable source of cytoplasmic male sterility (CMS), enhanced shelf life, high capsaicin content, earliness, high ASTA color value, high dry recovery percentage, are the major priority traits in the order in commercial breeding of chilli (Schreinemachers and Lin, 2023)

Varietal Achievements

Table 1. Improved cultivars of chilli developed by public sector in India

Cultivar	Special attributes	Source
CH 1	Hybrid derivative of MS12 × LLS. Tolerant ot fungal and viral diseases. Suitable for the processing industry.	PAU, Ludhiana
CH 3	Hybrid derivative of MS- 12 × S-253.	PAU, Ludhiana
CH 27	Nuclear male sterility-based hybrid. Moderately resistant to RKNs, anthracnose and leaf curl virus.	PAU, Ludhiana
CH 52	Hybrid between MS-13A × IS-261. Suitable for low tunnel cultivation. Moderately resistant to RKNs, anthracnose and leaf curl virus.	PAU, Ludhiana
Utkal Ava (BC 14-2)	Upward fruiting habits. Highly pungent. Suitable for both green and dry chilli purposes. Resistant to bacterial wilt.	OUAT, Orissa
Utkal Rasmi (BC 21-2)	Fruits are long, good in pungency, and deep red color on ripening. Suitable for dry chilli, resistant to bacterial wilt, and moderately resistant to fruit borers.	OUAT, Orissa
Arka Meghana	F_1 hybrid developed using CGMS line, tolerant to ChiVMV, thrips and suitable for dual-purpose market Yield: 5 tonnes of dry chilli per hectare.	IIHR, Bangalore
Arka Sweta	F_1 hybrid derived from a cross between IHR 3903 (CGMS line) × IHR 3315.Yield: 4.5 tonnes of dry chilli per hectare. It exhibits field tolerance to viral diseases.	IIHR, Bangalore
Arka Khyati	F_1 hybrid	IIHR, Bangalore
Arka Harita	F_1 hybrid derived from a cross between IHR 3905 (CGMS) × IHR 3312. Yield: 5.0-5.5 tonnes of dry chilli per hectare. It is tolerant to powdery mildew and viral diseases.	IIHR, Bangalore
Arka Tejasvi	High-yielding chilli F_1 hybrid. Yield: 3.0-3.5 tonnes of dry chilli per acre. Tolerant to chilli leaf curl virus.	IIHR, Bangalore
Arka Tanvi	F_1 hybrid suitable for green and dry chilli production. Yield: 3.0-3.5 tonnes of dry chilli per acre. It exhibits tolerance to chilli leaf curl virus.	IIHR, Bangalore

Arka Saanvi	F_1 hybrid suitable for green and dry chilli production. Yield: 3.0-3.5 tonnes of dry chilli per acre. It shows tolerance to chilli leaf curl virus.	IIHR, Bangalore
Arka Yashasvi	High-yielding F_1 hybrid. Yield: 3.0-3.5 tonnes of dry chilli per acre. It shows resistance to chilli leaf curl virus (ChLCV).	IIHR, Bangalore
Pant C-1	Derived from a cross between NP46A × Kandhari. The plants are tall and erect. Fruits are upright, highly pungent and tolerant to leaf curl and mosaic virus.	GBPUAT, Pantnagar
Pant C-2	Selection from a cross between NP46A × Kandhari. Tolerant to mosaic and leaf curl virus.	GBPUAT, Pantnagar
NP 46A	Selection from local, Puri Red. Dwarf and spreading type.	IARI, New Delhi
Pusa Jwala	Derived from a cross of NP 46A and puri red. Dwarf and spreading type. Tolerant to mosaic.	IARI, New Delhi
Pusa Sadabahar	Hybrid derivative of Pusa Jwala × IC 31339 (*C.Frutescense*). Cluster bearing habit (6-14 fruits). Resistant to leaf curl virus, TMV and CMV.	IARI, New Delhi
Kashi Tej (CCH-4)	CMS-based F_1 hybrid. High pungency. Early maturing. Dual-purpose (green and dry). Tolerant to thrips and anthracnose.	IIVR, Varanasi
Kashi Abha (VR-339)	Highly pungent. Tolerant to high and low temperature as well as thrips, mites, CLCV and anthracnose.	IIVR, Varanasi
Kashi Gaurav	Bushy. Yield 150 q/ha. Tolerant to mites, thrips and anthracnose.	IIVR, Varanasi
Kashi Ratna (CCH-12)	CMS based F_1 hybrid. Tolerant to thrips and anthracnose.	IIVR, Varanasi
Kashi Sinduri	Spreading type.Resistant to anthracnose.Average yield: 14 t/ha.	IIVR, Varanasi
Kashi Surkh (CCH-2)	CMS-based (CCA 4261) F_1 hybrid. Suitable for dual purpose (green and dry). Tolerant to leaf curl virus, mites and thrips.	IIVR, Varanasi
Punjab Lal	Hybrid derivative of Perennial x long red. Plants dwarf. Fruits are erect, medium-long, dark green, and highly pungent. Resistant CMV and leaf curl virus.	PAU, Ludhiana
Punjab Guchhedar	Chilli variety suited for processing.	PAU, Ludhiana
Punjab Sindhuri	It is suitable for fresh market and distant transportation. Average yield of red ripe fruits is 190 q/ ha.	PAU, Ludhiana
Punjab Tej	Highly pungent. Suitable for industrial use.	PAU, Ludhiana
Phule Jyoti	Developed through selection. Tolerant to bacterial wilt and fruit fly.	MPKV, Rahuri

Table 2. Chilli cultivars developed by private sector and being cultivated in India

Company	Leading hybrids
Mahyco	Yashaswini, Sierra, Navtej, MHCP 321, MHCP 310 (Teja), MHCP 318 (Tanaya), MHP 1(Tejaswini), MAHY 456, MAHY 453, Tanaya
Syngenta	HPH-12, Roshni, HPH-1900, HPH-490, HPH-117,Veera (70895), Ojaswi, HPH-22, HPH-21, HPH-1028, HPH-5531,HPH-2043, HPH-1048, HPH-4968(Bullet), Devsena 88
Namdhari	NS 1701, NS 1101, NS 211, NS 1072, NS 208, NS 227, NS 203, NS 238, NS 201, NS 202, NS 688
Nunhems	Armour, Indu, US 341, US 730, Sarpan, US 803, US 1003, US 702, Ojas, US 1081, Soldier, US 765, US 323, US 612, Indira Plus 1, US 1081, US 917, Laali, US 1195, Ujala, Nun 2074
Seminis	SHP 4884, Golden Hot, Sitara, Sitara Gold, Wonder Hot, Arunim, Mega Hot, Siam Hot
Ankur Seeds	ARCH 930, ARCH 2121, ARCH 2239, Amisha, Prajakta, Kajol, Kanchi, Sarangi, Monica
Indo American Hybrid Seeds	Indam 42, Indam 5, Indam Pruthvi, Indam Aruna, Indam Surya
US Agriseeds	SW 402, SW 407
VNR seeds	Shilpa (435/7), Pari, VNR 10, VNR 109, VNR 1262, VNR 1307, VNR 1366, VNR 145, VNR 1616, VNR 1786, VNR 1921, VNR 24, VNR 277, VNR 314, VNR 332 (Rani), VNR 377 (Krishna), Nupur, Ramba, Gundu, VNR-G-907, VNR-G-273 (Jyothi), Sahiba
East-West seed	Tejita 642, Ulka, Daiya 619, Venus 942, Rudra 031, Paleo, Hakone, Red Hot
Bejo Seethal	BSS 378, BSS275
Nirmal Seeds	NCH 1012, NCH 913, NCH 1526, NCH 12

Table 3. Sweet pepper cultivars developed by public and private sector in India

Public sector	
Variety/ Hybrid	**Characteristics**
Arka Mohini	Open pollinated variety (ICAR-IIHR). Determinate plant habit with dark green foliage, green blocky fruit, thick fleshed, 3-4lobbed, turns red on ripening
Arka Gaurav	Open pollinated variety released by IIHR, Bangalore. Indeterminate plant habit with green foliage, green oblong to blocky fruit, 3-4lobbed, turns yellow on ripening.
Arka Basant	Open pollinated variety released by IIHR, Bangalore. Indeterminate plant habit with yellowish green foliage, cream coloured conical fruit, 2-3 lobbed, turns orange red on ripening.
Arka Athulya F_1	F_1 hybrid with powdery mildew tolerance released by IIHR, Bangalore. Dark green blocky fruits with 3-4 lobes, turns red on ripening.
Pusa Deepti F_1	F_1 hybrid released from IARI regional station, Katrain. High yielding, light green, conical fruit. Turns red on ripening.

Variety/ Hybrid	Characteristics
Bokken F_1	Medium size, blocky fruit, good field holding capacity & shelf life, tolerance to Powdery Mildew & TSWV and suitable for Green & Red Fresh Harvest
Celin F_1	Strong vigorous plant with good foliage cover, perfect blocky fruit, good foliar disease tolerance, tolerance to PepMoV and suitable for Green & Red Fresh Harvest
Kaamos F_1	Medium size , blocky yellow fruit, good field holding capacity & shelf life, tolerance to Powdery Mildew, Nematodes & TSWV
Intruder F_1	Perfect blocky, green fruit, good foliar disease tolerance, thick fruit pericarp with good fruit weight and high yielding capacity
Indra F_1	Medium to big size fruit, good cold and heat set capacity, wider adaptability and high yielding,
Dollar F_1	Good Cold set & Heat Set, wider adaptability, high yielding capacity
Gracela F_1	Excellent fruit uniformity across the pickings, attractive smooth surface fruit with medium size, suitable for green Harvest
Orobelle F_1	Good yielding capacity, stable to grow in different conditions, suitable for open field and net house cultivation
Inspiration F_1	Medium to large-sized fruits, attractive, glossy red colour and a uniform blocky shape. With thick walls, offer a long shelf life and are ideal for long-distance shipping. Best suited for protected cultivation
Bachata F_1	High-yielding yellow blocky pepper for protected cultivation, large sized blocky fruits are thick walled ideal for long distance shipping
Pasarella F_1	Compact plant with elongated uniform fruits, recommended for green and red harvest under green house /shade net condition

All the coloured bell pepper F_1 hybrids marketed in India are imported. Popularly grown commercial hybrids include Inspiration , Bachata, Pasarella, Intruder, Bokken, Kaamos etc.

Commercial Hybrid Seed Production Practices

While chilli is typically classified as a self-pollinated crop, cross-pollination rates ranging from 60-90% have been reported, possibly attributable to the exerted nature of the stigma. Consequently, to uphold the purity of parental lines, a recommended isolation distance of at least 400-500 meters is advised. The steps involving the development of F_1 hybrids in *Capsicum* are 1. Production of inbred lines 2. Testing of combining ability and 3. Hybrid seed production (Either through emasculation & pollination or, use of male sterility).

Male Sterile Systems in *Capsicum*

i) **Genetic Male Sterility:** The inheritance of the genetic male sterile trait typically follows a single recessive gene, denoted as *ms*. While the utilization of Genetic Male Sterility (GMS) holds promise for hybrid seed production, its efficacy is hindered by challenges in establishing and maintaining a male sterile plant population. To address this limitation, a cross is conducted between a heterozygous fertile plant and a male sterile plant, resulting in progeny that segregates in a 1:1 ratio for the sterile and fertile traits. Precise removal of the 50% male fertile plants is imperative to prevent contamination. The integration of linked molecular markers for the male sterile trait offers a promising solution by enabling the early identification of sterile plants, thereby enhancing the efficiency and accuracy of male sterile plant selection in hybrid seed production.

ii) **Cytoplasmic male sterility:** Cytoplasmic male sterility (CMS) presents an alternative method for hybrid seed production, offering the advantage of generating 100% sterile plants, thereby eliminating the risk of contamination. Sterility arises from the interaction between nuclear and cytoplasmic factors. In chili, stable cytoplasmic-nuclear male sterile lines (CMS lines) are available, serving as the female parent in hybrid seed production. Utilization of such lines alleviates the substantial costs associated with manual emasculation, given the sterility of the male part (characterized by constricted anthers with minimal pollen). Consequently, employing CMS lines significantly reduces the production costs of hybrid seeds. The CMS line, also known as the A line, is maintained through crosses with pollen from the maintainer line (B line), ensuring the generation of 100% male sterile seeds, unlike in GMS lines. Seedlings resulting from these crosses can be transplanted, with four female rows alongside one male row comprising the selected restorer line (R line) in the hybrid seed production field. Given the significant natural cross-pollination observed in chili, manual pollination is unnecessary; however, for optimal seed yield ranging between 300-350 kg/ha, hand pollination is recommended. Further systematic research on honeybee activity's role in enhancing natural seed yield in chili utilizing CMS lines is warranted

The development of the CMS line (A line) involves a backcrossing strategy wherein a chosen maintainer line (B line) is crossed onto an existing superior inbred line. This breeding approach results in the creation of a pair of A and B lines within the new genetic background. The restorer allele (*Rf*) is introduced either through introgression into a designated male parent (restorer breeding) or directly utilizing a male parent for hybrid seed production on CMS, provided the restorer (*Rf*) gene is already present in a homozygous state in the male parent. The advantage of the Cytoplasmic Genetic Male Sterility (CGMS) system lies in its ability to generate a population of sterile plants wherein all offspring are sterile. CGMS represents the most commonly employed male sterile system for commercial hybrid production in chili, operating on a three-line breeding scheme involving the A line (male sterile; S-rr), B line (maintainer; N-rr), and either a C or R line (restorer; S or N-RR). Notably, at IIHR, Bangalore, stable CGMS lines have been developed, facilitating the creation of several chili hybrids for commercial cultivation, including Arka Meghana, Arka Sweta, Arka Harita, Arka Yashasvi, Arka Taanvi, Arka Saanvi, Arka Gagan, Arka Tejasvi, Arka Nihira, and Arka Dhriti.

Table 4: Markers reported for non-allelic nuclear male sterile genes in pepper

gene	Marker type	Chromosome location	Reference
ms3	SNP based CAPS	Chromosome 1	Naresh Ponnam *et al.*, 2018
ms10	SSR	Chromosome 1	Aulakh *et al.*, 2016
ms8	SCAR	Chromosome 4	Bartoszewski *et al.*, 2012
ms_w	SNP based CAPS	Chromosome 5	Naresh Ponnam *et al.*, 2018
ms1	AFLP	Not reported sequence unavailable)	Lee *et al.*, 2010a
	HRM	Chromosome 5	Jeong *et al.*, 2018
ms3	AFLP- CAPS	Not reported (sequence not available)	Lee *et al.*, 2010b

Table 5. Sterile cytoplasm specific markers

Marker	Marker type	Primers	Reference
atp6 SCAR	SCAR	F- AGTCCACTTGAACAATTTGAAATAATC R- GTTCCGTACTTTACTTACGAGC	Kim and Kim 2005
orf456-SCAR	SCAR	F- ATGCCCAAAAGTCCCATGTA R- TTACTCGGTTGCTCCATTGTTT	Kim *et al.* (2007)

Table 6. *Restorer of Fertility (Rf)* gene specific markers

Marker	Marker type	Primers	Restriction enzyme	Reference
CaRf648	SNP based CAPS	F-CCTTATGAATGAGATGAAAC R-CATCCATTATCGCATTGTCG	HindIII	Ortega *et al.*, 2020
ColMod1-CAPS	CAPS	F- CCACCCCAAACGTAAGGGATC R-CAACTTAGCCAATACAATCCCAC	*Bgl*II	Jo *et al.*, 2016

Package of Practices for Seed Production in Chilli

Raising chilli seed crops requires specialized knowledge in crop management, including preventing outcrossing, curing, processing, and storage of seeds. Special attention is given to crop inspections, removal of off-types, disease control, harvesting, extraction, and processing of seeds.

Isolation: To maintain varietal purity, the seed field should be isolated from other varieties and non-conforming fields by distances of 400 and 200 meters for foundation and certified seed production, respectively. Hot pepper fields should also be separated from sweet pepper fields and vice versa due to their cross-compatibility.

Figure 1. Hybrid seed production in chilli **A.** Insect proof net house production, **B.** Male fertile flower (male parent), **C.** Male sterile (Female parent), **D.** Anthers collection from male parent, **E.** Pollen powder, **F and G,** Ring pollination and **H.** Set fruit

Field Standards

General Requirements

Contaminants	Minimum distance (meters)	
	Foundation	Certified
Fields of other varieties	50	25
Fields of the same variety not conforming to varietal purity requirements for certification	50	25

Specific Requirements

Factor	Maximum permitted (%)*	
	Foundation	Certified
Off types	0.10	0.20
** Plants affected by seed borne diseases like Leaf blight (*Alternaria solani* Sorauer), Anthracnose (fruit rot and die-back) (*Colletotrichum capsici* (Syd.) Butter and Bisby	0.10	0.50

Land Requirement: Specific land requirements regarding previous crops are not mandated under Indian Minimum Seed Certification Standards. However, it is suggested to avoid raising chili crops on the same land in consecutive seasons to prevent volunteer plant growth.

Field Inspection and Rouging: Inspections are conducted at various growth stages to ensure genetic purity. Off-type plants are removed, particularly before flowering, at full bloom and fruiting stage, and just before the first fruit picking to avoid mechanical admixtures and outcrossing.

Harvesting and Seed Extraction: Chilli fruits are harvested for seed when fully ripe and red. Common seed extraction methods include sun drying followed by crushing and winnowing or manual extraction by squeezing fruits. Mechanical extraction machines are also used to retain seed quality.

Seed Drying and Storage: Freshly extracted seeds have high moisture content and must be quickly cleaned and dried to below 8 percent moisture for proper storage. Drying methods include sun drying or using screen bottom trays exposed to sunlight. Proper labeling is crucial for maintaining seed identity during storage, which can be extended by storing seeds under low temperature and humidity conditions. Seeds should have a minimum germination rate of 60 percent and can be stored in cloth bags or sealed containers.

Seed Processing: Seed processing for chilli seeds involves several essential steps to ensure the quality and purity of the seeds. Initially, the harvested seeds undergo cleaning to remove any debris, plant materials, and other impurities. This is typically done using a seed cleaner or gravity separator. Following cleaning, the seeds are dried to reduce their moisture content to a suitable level for storage, utilizing equipment such as a seed dryer or drying floor. Once dried, the seeds are separated from any remaining fruit pulp or contaminants through a process of separation, which can be achieved using a seed separator or winnower. Grading then takes place to sort the seeds based on size, weight, or other characteristics, often facilitated by a seed grader or sorter. After grading, the seeds may undergo treatment with fungicides, insecticides, or other protective agents to safeguard against pests and diseases, applied using a seed treater or coating machine. Subsequently, the seeds are packaged in appropriate containers for storage or distribution, a task performed by a seed packaging machine or bagging equipment. Finally, the seed packets are labeled with essential information such as variety, processing date, and planting instructions, typically achieved with a labeling machine or printer. Throughout these processes, strict adherence to cleanliness, precision, and quality control is vital to ensure the production of high-quality chilli seeds.

Seed Standards

Factors	Minimum percentage allowed	
	Foundation	Certified
Pure Seed (minimum)	98.0%	98.0%
Inert matter (maximum)	2.0%	2.0%
Other crops seeds (maximum)	5/kg	10/kg
Weed seeds (maximum)	None	None
Germination (minimum)	60%	60%
Moisture (maximum)	8.0%	8.0%
For vapour-proof containers (maximum)	6.0%	6.0%

Seed Yield: 300-350 kg/ ha; 30-35 kg/ 10 guntas; 10-15 g/ plant

Precautions for Quality Seed Production

- To ensure purity, it is recommended to cultivate both male and female parents within a controlled environment such as a nethouse or polyhouse.
- The optimal number of hybrid fruits per plant varies: 5-10 in bell pepper and 50-70 in chilli.
- After completing the hybridization program, all non-crossed (untagged) fruits resulting from natural crossing on female plants should be removed. This practice promotes the robust development of crossed fruits and seeds.
- In the context of CMS/GMS-based hybrid seed production, the removal of non-crossed fruits is unnecessary, given that all fruits developed on male sterile plants will be crossed, provided recommended isolation distances are adhered to.
- Harvesting of crossed fruits should occur when chili and bell pepper fruits reach maturity, typically between 35-50 days after pollination, indicated by red ripe or fully colored fruit.
- Prior to harvesting crossed fruits, it is essential to remove open-pollinated (non-hybrid) fruits to minimize the risk of contamination. Only tagged fruits should be harvested to ensure purity.
- Seed extraction and packaging involve drying the harvested ripe fruits, followed by seed separation through maceration (for commercial scale) or longitudinal bifurcation (for experimental scale). Seeds should be dried to an 8% moisture level before packing in moisture-proof materials. Cleaning seeds with a density gradient seed cleaner is recommended.

Table 7: DUS characteristics in chilli and capsicum

S. No.	Characteristics
1	Seedling: Anthocyanin colouration of hypocotyl
2	Plant: Habit
3	Plant: Length of main stem
4	Plant: Length of first internode (on primary branches in cm)
5	Plant: Anthocyanin colouration of nodes
6	Plant: Intensity of anthocyanin colouration of nodes
7	Plant: Height (cm)
8	Plant: Spread (cm) (average of distance between the extremes at the widest points taken in two directions)
9	Stem: Pubescence (hairiness)
10	Stem: Intensity of pubescence
11	Stem: Shape
12	Leaf: Length of blade (measured from leaf base to leaf tip in cm)
13	Leaf: Width of blade (measured on the widest part of the leaf in cm)
14	Leaf: Colour
15	Leaf: Intensity of green colour
16	Leaf: Intensity of purple colour
17	Leaf: Shape
18	Leaf: Undulation of margin
19	Leaf: Pubescence (hairiness)
20	Leaf: Intensity of pubescence (hairiness)
21	Flower: Petal colour
22	Flower: Anther colour
23	Flower: Days to 50% flowering (from the date of sowing)
24	Flower/ Fruit: Orientation
25	Fruit: Bearing habit (no. of fruits/ node)
26	Fruit: Colour (at mature unripe stage)
27	Fruit: Intensity of colour (at mature unripe stage)
28	Fruit: Length (cm)
29	Fruit: Diameter (fruits measured at the widest point in cm)
30	Fruit: Shape in longitudinal section
31	Fruit: Curvature
32	Fruit: Curvature intensity
33	Fruit: Neck at basal end
34	Fruit: Cross sectional corrugation (at level of placenta)
35	Fruit: Sinuation of pericarp
36	Fruit: Texture of surface

37	Fruit: Colour (at ripe maturity)
38	Fruit: Intensity of colour (at maturity)
39	Fruit: Color transition
40	Fruit: Glossiness
41	Fruit: Shape at the base
42	Fruit: Shape of apex
43	Fruit: Number of Locules/ lobes (bell pepper)
44	Fruit: Pericarp thickness (at physiological mature stage)
45	Fruit: Stalk Length (cm)
46	Fruit: Calyx Cover
47	Fruit: Calyx Margin
48	Fruit: Calyx Constriction
49	Fruit: Pedicel attachment
50	Fruit: Blossom end appendage
51	Fruit: Days to 50% ripening (from the date of sowing)
52	Seed: 1000 seed weight (g)
53	Seed: Recovery (%)
54	Seed: Colour
	Specialty trait
55	Male sterility

References

Aguilar-Meléndez, A., Morrell, P. L., Roose, M. L. and Kim, S. C., 2009. Genetic diversity and structure in semiwild and domesticated chiles (*Capsicum annuum*; Solanaceae) from Mexico. American Journal of Botany, 96(6):1190-1202.

Aulakh, P. K., Dhaliwal, M. S., Jindal, S. K., Schafleitner, R., and Singh, K., 2016. Mapping of male sterility gene ms10 in chilli pepper (*Capsicum annuum* L.). Plant Breed. 135, 531–535.

Bartoszewski, G., Waszczak, C., Gawroński, P., Stępień, I., Bolibok-Brągoszewska, H., Palloix, A., Lefebvre, V., Korzeniewska, A. and Niemirowicz-Szczytt, K., 2012. Mapping of the ms8 male sterility gene in sweet pepper (*Capsicum annuum* L.) on the chromosome P4 using PCR-based markers useful for breeding programmes. Euphytica, 186: 453-461.

Carrizo García, C., Barfuss, M.H., Sehr, E.M., Barboza, G.E., Samuel, R., Moscone, E.A. and Ehrendorfer, F., 2016. Phylogenetic relationships, diversification and expansion of chili peppers (Capsicum, Solanaceae). Annals of Botany, 118(1): 35-51.

Hernández-Verdugo, S., Luna-Reyes, R. and Oyama, K., 2001. Genetic structure and differentiation of wild and domesticated populations of *Capsicum annuum* (Solanaceae) from Mexico. Plant Systematics and Evolution, 226: 129-142.

Jeong, K., Choi, D. and Lee, J., 2018. Fine mapping of the genic male-sterile ms 1 gene in Capsicum annuum L. Theoretical and Applied Genetics, 131:183-191.

Jo, Y.D., Ha, Y., Lee, J.H., Park, M., Bergsma, A.C., Choi, H.I., Goritschnig, S., Kloosterman, B., van Dijk, P.J., Choi, D. and Kang, B.C., 2016. Fine mapping of Restorer-of-fertility in pepper (*Capsicum annuum* L.) identified a candidate gene encoding a pentatricopeptide repeat (PPR)-containing protein. Theoretical and Applied Genetics, 129: 2003-2017.

Kim, D.H. and Kim, B.D., 2005. Development of SCAR markers for early identification of cytoplasmic male sterility genotype in chili pepper (*Capsicum annuum* L.). Molecules and Cells, 20(3): 416-422.

Kim, D.H., Kang, J.G. and Kim, B.D., 2007. Isolation and characterization of the cytoplasmic male sterility-associated orf456 gene of chili pepper (*Capsicum annuum* L.). Plant Molecular Biology, 63: 519-532.

Lee, J., Yoon, J.B., Han, J.H., Lee, W.P., Do, J.W., Ryu, H., Kim, S.H. and Park, H.G., 2010a. A codominant SCAR marker linked to the genic male sterility gene (ms1) in chili pepper (*Capsicum annuum*). Plant Breeding, 129(1): 35-38.

Lee, J., Yoon, J.B., Han, J.H., Lee, W.P., Kim, S.H. and Park, H.G., 2010b. Three AFLP markers tightly linked to the genic male sterility ms 3 gene in chili pepper (*Capsicum annuum L.*) and conversion to a CAPS marker. Euphytica, 173: 55-61.

Naresh Ponnam, Shih-wen Lin, Chen-yu Lin, Yen-wei Wang, Roland Schafleitner, Andrzej Kilian and Sanjeet Kumar, 2018. Molecular Markers Associated to Two Non-allelic Genic Male Sterility Genes in Peppers (*Capsicum annuum* L.). Frontiers in Plant Science, 9: 1343.

Ortega, F.A., Barchenger, D.W., Wei, B. and Bosland, P.W., 2020. Development of a genotype-specific molecular marker associated with restoration-of-fertility (Rf) in chile pepper (*Capsicum annuum*). Euphytica, 216: 1-10.

Reddy, M. K., Srivastava, A., Kumar, S., Kumar, R., Chawda, N., Ebert, A. W. and Vishwakarma, M., 2014. Chilli (*Capsicum annuum* L.) breeding in India: an overview. SABRAO Journal of Breeding and Genetics, 46(2):160-173.

4

Okra

M. Pitchaimuthu

ICAR-Indian Institute of Horticultural Research, Hesaraghatta Lake Bengaluru-560089, Karnataka

Introduction

Okra [(*Abelmoschus esculentus* (L.) Moench] is one of the traditional vegetable crop cultivated in tropical, subtropical and warmer regions of the world. Okra fruits are rich in calcium (90 mg/100 g fresh weight) and provide valuable supplementary items in the tropical diet which is starchy in nature lacking calcium and iron. The root and stem are used for clearing cane juice in preparation of jaggery or 'gur'. The high iodine content of fruits helps prevent goiter, while leaves are used in inflammation and dysentery. It has yet multiple uses. The dry seed contains 13-22 % good edible oil and 20-24 % protein. The oil is used in soap, cosmetic industry and as vanaspati, while protein is used for fortified feed preparations. The crushed seed is fed to cattle for increasing the milk production and the fibre is utilized in jute, textile and paper industries. India ranks first in the world with 7.16 million tonnes of production 5.49 lakh hectares of area. Gujarat is the leading okra-producing state which produces around 1147.66 thousand tonnes from an area of 93.95 thousand hectares followed by West Bengal. Fresh fruit of okra is exported to UAE, Sri Lanka, the Netherlands, Bangladesh, Malaysia, Nepal, UK, Saudi Arabia and Qatar. The national average productivity is very low (11.70 t/ha), that the crop is prone to damage by major biotic stresses like fungus, nematodes and viruses like Yellow Vein Mosaic Virus and Enation Leaf Curl Virus. The yield potential of okra is low; however, it could be improved through hybridization. Marked heterosis has been reported in okra for yield and its components.

Floral Biology and Pollination Behaviour of Okra

The bisexual (hermaphrodite) flowers open in the morning between 7.30 – 10.00 am. The stigma becomes receptive 6 hours before anthesis and so remains 6 hours after anthesis. The anther dehiscence occurs between 5-7 am. Okra is an often-cross-pollinated crop with cross-pollination ranging from 4-42 %. Honeybees and several other wild bees are responsible for cross-pollination. An isolation distance of 1000 meter and 500 meter is recommended for foundation and certified seed production, respectively.

Techniques of Hand Emasculation and Pollination: Hand Emasculation

As the flowers in okra are large and solitary, it is comparatively easier to emasculate and cross-pollinate them. A day prior to opening, the flower buds are emasculated, preferably in the afternoon by scrapping off the anthers from the staminal column with the help of the blunt edge of a scalpel. As the floral organs are mucilaginous there are chances of anthers adhering to the staminal column or stigma while removing them, great care is needed in emasculation so that no anther is left sticking to the floral parts. Immediately after emasculation, the flower bud should be covered with butter paper bags. Incase of the pollen parent, the flower buds which are supposed to open in the next day should be tied at the apex of the corolla with a thread or adhesive cellophane tape to prevent their opening.

Pollination

For pollination, the pollen from the freshly opened but previously bagged or tied flower from the pollen parent is dusted directly onto the stigma of the emasculated flowers of the female parent. The pollen can also be applied with the help of a camel hairbrush. For best results, the pollination should be done between 7.30 to 10.30 am. Emasculation and pollination can be continued for 40-45 days after sowing and it requires 5 labourers per day per acre.

Use of Male Sterility for Hybrid Seed Production

Presently, Genic Male Sterile (GMS) system is being explored in hybrid seed production. A field design in which alternate planting of two rows along with male sterile and fertile plants was done keeping a ratio of 1:1 and maintaining a plant population of 4,166 male sterile plants per hectare, gave hybrid seed yield of 5.66 quintals per hectare, which was more by 52.56 percent over with the fertile control. By using male sterile plants and hand pollination, it takes 2 hrs. and 53 minutes to pollinate 274 flowers sufficient to produce 1 kg of seed. By hand emasculation and pollination of 274 perfect flowers, it takes 9 hours and 37 minutes. Thus approximately 70 percent saving in time and manual labour can be achieved in producing 1 kg of hybrid seeds using male sterile lines in okra.

Hybrid seed production in okra is time-Commercial consuming and cumbersome process involving hand emasculation and pollination. Stable monogenic recessive Genic Male Sterility (GMS) has been developed through mutation breeding by gamma radiation at ICAR-IIHR, Bengaluru (Dutta, 1971) and this genic male sterile gene was incorporated into many different genetic backgrounds like Arka Anamika (GMS-1), Arka Abhay (GMS-2), multiple ribs line (GMS-3) and Dark green segment IIHR-253 (GMS-4) and GMS based F_1 hybrids were developed (Pitchaimuthu and Dutta, 2002).

Further, in comparison to conventional seed producton, there is 70% of reduction in time and labour requirements when GMS system is used. Hybrid seed production in okra is labour intensive. It requires 180 man-days per acre to produce hybrid seeds by emasculation and pollination. Further use of male sterile lines protects okra labour force from physical hand injury and skin irritation, thereby increasing their efficiency and reducing their physical drudgery (Pitchaimuthu *et al.* 2012).

Adoption of okra GMS-based F_1 hybrid – Arka Nikita by the growers, gave an average net return of Rs. 40,000 per hectare due to 25 % increased yield over other commercial hybrids and Rs. 85000 net returns from commercial hybrid seed production.

Genetic male sterility systems have been utilized for commercial hybrid production in okra. The seed and pollen parents are grown in 4:1 ratio. However, to maintain the good plant population in female rows it is suggested that seed parent should be sown with three to four seeds per hill and subsequent thinning of the excess plants after germination. The male sterile line is maintained in heterozygous form by crossing with the maintainer line under adequate isolation distance or under cage of 40 mesh nylon net. The yield potential of the GMS-based okra hybrid Arka Nikita (25 t/ha) will bring up the national productivity which is low (11.6 t/ha) at present. The high yield potential of the GMS hybrid will reduce the cost of production and increase profitability with high B:C ratio to the farmers in the country (Pitchaimuthu *et al.* 2012).

Use of GMS system in okra hybrid seed production at commercial level

Arka Nikita F_1 hybrid

GMS based Hybrid seed production

Conventional method of seed production

Removal of antthridial cone without finger cap Removal of antthridial cone with finger cap

Stigma and style afer emascualation without cap with cap

Stigma and style after emasculation

Hand Emasculation and pollination in Okra

Hand pollination techniques

Holding pollen grain on thumb nail

Dusting of pollen grain on the stigma

Putting proper label of hybrid combination pollination

Covering the butter paper cover after

Day before anthesis

Day of anthesis

Pollination techniques in Okra

Table 1. Advantages and disadvantages of GMS-$_1$ seed based hybird production in okra

Sl. No.	Parameters	Advantages	Disadvantages
1	Hand emasculation	Not required	Required
2	Produce 1 kg of hybrid seeds No. of flowers to be crossed	274 2 hrs 53 minutes	274 9 hrs 37 minutes
3	Seed rate (kg /ha)	6 kgs	4 kgs
4	Spacing (cm)	60 × 15	60 × 30
5	Removal of fertile plants	Required	Not required
6	Cost-benefit ratio	1: 4.36	1: 2.68

Table. 2. Cost of hybrid seed production in Okra

Sl. No.	Particulars	Geneic Male Sterile line	Normal fertile line (Rs/ha)
1	Land preparation	3000.00	3000.00
2	Seeds and sowing	2350.00	2350.00
3	FYM + Fertilizer and application cost	16535.00	16535.00
4	Intercultural Operations	2800.00	2800.00
5	Emasculation and Pollination	22230.00 *	44460.00
6	Plant protection	6200.00	6200.00
7	Irrigation	7500.00	7500.00
8	Harvesting and cleaning	1200.00	1200.00
9	Total cost cultivation (Rs)	61815.00	83695.00

GMS-based hybrid seed production achievement : 70 % saving in Time and Manual Labour

Economics of Hybrid Seed Production Using Gms Line in Okra (per hectare)

Seed yield (Q/ha)	:	5.99	5.00
Gross return (Rs)	:	2,69,550	2,25,000
Procurement Price (Rs)	:	450/kg	
Benefit-Cost ratio (Rs)	:	4.36	2.68

*No emasculation only pollination

Climate: Okra requires a long, warm and humid growing condition. It can be successfully grown in hot humid areas, but is sensitive to frost and extremely low temperature. For normal growth and development, a temperature between 24 ^{0}C and 28 ^{0}C is preferred. For seed germination optimum soil moisture and a temperature between 25 ^{0}C and 35 ^{0}C is needed. However, the fastest germination is observed at 35 ^{0}C. Beyond this range the germination will be delayed.

Soil: It is grown on sandy to clay soils but due to its well-developed tap root system, relatively light, well-drained, rich soils are ideal. As such, loose, friable, well-manured loam soils are desirable. A pH of 6.0-6.8 is ideally suited for commercial production as well as seed production of okra.

Land preparation: Bring the soil to a fine tilth after 2-3 ploughings and harrowing. Level the land with a leveller. Incorporate well-rotten FYM @ 25 tonnes per hectare to improve the soil texture and aeration. Prepare ridges and furrows spaced 60 cm apart during *Kharif* and 45 cm apart during the summer season.

Table 3. Manures and Fertilizers

Source	**Per hectare**
Organic manures	25 tonnes FYM
Nitrogen	100 kg (500 kg Ammonium Sulphate)
Phosphorus	50 kg (312 kg Single Super Phosphate)
Potassium	50 kg (83 kg Muriate of Potash)

Application: Before sowing, incorporate 25 tonnes of FYM per hectare. Open up a deep narrow furrow on one side of each sowing ridge. Apply the fertilizer mixture containing half dose nitrogen, full dose of phosphorous and potash in these furrows and cover the fertilizers with soil and then irrigate. Application of neem cake @ 250 kg/ha will reduce the incidence of sucking insect pests during the early stage of crop growth

Table 4. Sowing time

Southern plains:	1. June-July
	2. September-October
	3. February-March
Northern and Western plains:	1. July-August
	2. February-March
Eastern plains:	1. May-June
	2. February-March
Hilly Areas:	April- June.

Table 5. Hybrids Released from Public Sectors

SL No	Name of the hybrid	Sources	Remarks
1.	Hybrid-7 and Hybrid-8	TNAU, Coimbatore	Purple colour fruit, smooth, tender, good keeping and cooking qualities. It is tolerant to yellow vein mosaic virus.
2.	GOH-3, GOH-4	GAU, Anand	Green, smooth tender, good keeping and cooking qualities. It is tolerant to yellow vein mosaic virus.
3.	Shitala Upahar	IIVR, Varanasi	Green, smooth, medium fruit, yield 20 t/ha, recommended for all okra growing states
4.	Shitala Jyothi	IIVR, Varanasi	Green, smooth, medium fruit yield 19 t/ ha, recommended for Rajasthan, Gujarat, Haryana, Delhi, Orissa, Chattishgarh and A.P. Tolerant to YVMV and OLCV
5.	Kashi Bhairav	IIVR, Varanasi	Light Green, smooth, medium fruit, yields 18 t/ha, recommended for all okra growing states.
6.	Kashi Rupa	IIVR, Varanasi	Lush Green, smooth, medium fruit, yield 18-20 t/ha, recommended for all okra growing states
7.	Arka Nikita	IIHR, Bengaluru	Early maturity (first at harvest 45days DAS),dark green, smooth fruit, 3-4 productive branches, medium fruit, yield 21-24 t/ha and tolerant to YVMV . It is first commercial GMS based hybrid in the country.

Table 6. Okra F_1 hybrids released from Private Sector

SL No	Name of the hybrid	Sources	Remarks
1.	Adhunik and Panchali	Century Seeds	Green fruits, medium yielder
2.	Varsha and Vijay	Indo American Hybrids Seeds	Green fruits
3.	Mahyco-No-10	Mahyco Seeds	Short, green fruits and tolerant to YVMV
4.	Nath Shoba	Nath Seeds	Light green fruit
5.	Supriya	Sun Seeds	Green fruits
6.	Nirmal-15, Nirmal-101(Nisha), Nirmal-147 and Nirmal-303	Nirmal Seeds	Green fruits, tender, and tolerant to YVMV
7.	NS-531, NS-532, OKH-803, OKH-808, OKH-812, OKH-814, OKH-818, OKH-833	Namdhari Seeds	Medium tall plant, med. to long fruits dark green fruits, Good yield and very good cooking quality and resistant to YVMV and ELCV

8.	Syngenta-152	Syngenta Seeds	Tall plant, good branching, smooth, tender dark green fruits and tolerant to YVMV
9.	US-7109	US Agri Seed Works	Medium plant and branching, Dark green colour fruits, smooth and tender and tolerant to YVMV

Spacing: A planting distance of 60 cm × 30 cm, accommodating 55,000 plants/ha is recommended for *kharif* season while 45 cm × 30 cm accommodating 75,000 plants/ha is recommended in the summer season.

Seed rate: 8-10 kg/hectare.

Sowing of seeds: Okra gives little success on transplanting and thus seed is sown directly in the soil by seed drill, hand dibbling or behind the plough. Sowing on ridges ensures proper germination, reduces water requirement during spring-summer and helps in drainage during the rainy season. Soaking seeds in 0.2 % Carbendazim solution overnight helps hasten germination and protects seedlings from Fusarium wilt. Sufficient soil moisture and temperature around 30° C help in quick and uniform germination. Sowing in moist soil is preferred over application of irrigation after sowing.

Irrigation: Irrigate the furrow lightly soon after the seed sowing to ensure good germination. Irrigate subsequently at an interval of 3-4 days depending upon the soil and weather conditions. If the temperature goes beyond 40 °C, frequent light irrigations are recommended to help proper fruiting. Drip irrigation saves around 85 % water though it is not been adopted at a commercial scale in okra. If drip irrigation is followed, then double row system of sowing has to be adopted.

Thinning of plants: Thin out the closely germinated plants at first true leaf stage.

Weed control: Proper weed management in okra could save up to 90 percent crop losses due to weeds. A total of 3-4 weeding starting from 20 days after sowing are required till the crop covers the soil surface.

Top dressing

Apply the remaining half of the recommend dose of nitrogen at the base of each channel 30 days after sowing, followed by earthing-up operation.

Rouging

Rouging of the seed crops should be begin with uprooting and destroying of YVMV infected plants after they are noticed. This should be continued up to the three-fruit stage. Subsequent roughing of off-types and wild species of

Abelomoschous sp. should be done prior to flowering. The off-type plants are easily distinguishable based on their plant height, leaf and stem characteristics, pigmentation, flower size and shape of the fruit etc.

Insect Pest and Nematode Management

- Apply good quality oiled neem cake containing 8% oil @ 250 kg/ha around the plant base or to ridges immediately after germination and cover with soil. This will reduce the incidence of nematodes and many insects.
- If the soil is sick with root-knot nematodes, apply the above dose of neem cake or Pongamia cake and also apply *Trichoderma* enriched FYM to soil before land preparation. For this purpose, mix 2-3 kg of *Trichoderma* with 1 ton of FYM, cover with polythene or dry grass and keep it under shade for one week and then apply to soil. Apply 5 tons of FYM thus enriched with the bio-pesticide.
- In addition to the above, treat the seeds with bio-pesticide *Pseudomonas fluorescens* @ 10 g/kg seed before sowing to manage nematodes.
- Spray imidacloprid (0.3 ml/l) or Thiamethoxam (0.3 g/l) 10 days after sowing to prevent plant hoppers. Do not repeat this spray thereafter, as they will leave harmful residues in harvested fruits.
- Repeat neem cake application after 30 days of sowing at the time of dressing and spray neem seed powder extract 4 % (for this purpose soak 4 kg of pulverized neem seed powder in 10-15 litres of water overnight and then filter through a fine cloth, makeup the volume to 100 litres and spray. One hectare requires 40 kg of neem seed powder and 1000 litres of spray fluid for one spray) or neem soap 1% at 10 days interval after flowering. Spray the lower leaf surface thoroughly where hoppers and mites are generally found. This will take care of petiole maggot, okra fruit borer and moderate levels of hopper and red spider mite incidence. Spraying of botanicals should be preferably done in the evening.
- If leaf miner incidence is noticed, remove all infested leaves and spray neem seed powder extract 4% or neem soap 1%.
- To control fruit borer incidence spray indoxacarb @ 0.75 ml/l at the time of flowering.
- To control okra fruit borer as an alternative to synthetic insecticides spray carbaryl (3 g/l). However, maintain a waiting period of 4 days after spray before harvest.
- If hopper incidence increases, then only spray Acephate (1 ml/l).

- If red spider mites incidence is noticed, first remove infected leaves and burn them. Then spray neem soap 1% or dicofol (2.5 ml/l) or wettable sulphur (3 g/ litre). However, care should be taken to spray lower surface of the leaves where mites are generally found.
- Maintain a waiting period of 7 days after synthetic pesticide spray like Acephate, indoxacarb and dicofol (for commercial crop).

Disease Management

- **Fusarium wilt**: It is caused by fungus *Fusarium oxysporum* f. sp. *vasinfectim*. Symptoms: Clearing of the veinlets and chlorosis of the leaf, soon petiole and leaf droop and wilt. In advanced cases, if the roots and basal stem are split open and examined, dark-brown or black discoloration of the vascular tissue may be seen. Drench the soil with 0.2 % Carbendazim solution to control fusarium wilt.
- Spray Mancozeb @ 2 g/l or Chlorothalonil @ 2 g/l of water to control Alternaria blight.
- Spray Dinocap @ 0.5 ml/l or wettable sulphur @ 2 g/l or Systhane @1 g/l or Contaf @ 1 ml/l of water to control powdery mildew.
- **Yellow Vein Mosaic Virus (YVMV)**: The disease is characterized by yellowing of the entire network of veins in the leaf blade. In severe infection, younger leaves turn yellow and become reduced in sizeand plant remains highly stunted. In severe cases, fruits become yellow and unfit for marketing and consumption. It is transmitted by vector whitefly *Bemisia tabaci*. To control this disease the following steps can be adopted:
    - Grow boarder crop with two rows of maize or jowar.
    - Spray systemic insecticides such as Acephate @ 1.5 g/l of water.
    - Remove the yellow vein mosaic virus affected plants at the early stage and burnt it.

Harvesting and Threshing: Dried fruits should be harvested (35 days after pollination), varieties with angular fruits, which open along sutures, should be harvested promptly to avoid shattering. Harvest the okra fruits when splits start appearing along the sutures. Cure the fruits for a week. Thresh the fruits and extract the seeds. Dry the seeds in sunlight for 2-3 days to get at least 10 % moisture.

Yield: Seed yield 7.5 to 10 quintals/ha depending upon variety/hybrid.

Seed standards of okra seed production: The following seed standards should be maintained in okra seed production.

Table 7. Seed standards in Okra

Standard	Foundation (%)	Certified (%)
Pure seed (Minimum)	99.0	99.0
Inert matter (Maximum)	1.0	1.0
Other crop seed (Maximum)	None	None
Total Weed seed (Maximum)	None	None
Objectionable weed seed (Maximum)	None	None
Germination (%) (Minimum)	65	65
Moisture Content (%) (Maximum)	10	10

Plant Protection in Okra Seed Production

Disease/insect pest	Control measures
Downey mildew (*Pseudoporonospora cubensis)*	Ridomil (0.2%) spray at 15 days interval
Powdery mildew (*Sphaerothece fulginiea*)	Bavistin (0.2%) sprays at 15 days interval Karathene (0.05%) spray at 10 days interval
Anthracnose	Foltof (0.2%) spray at 10 days interval
Alternaria blight	Javach (0.2%) spray at 10 days interval
Fusarium wilt	Drench the soil with 0.2% Bavistin
Phytophthora rot	Spray Cartan (0.2%) at the time of fruit ripening
Aphid/jassids/thrips	Spray Demacron (0.05%) or Rocor (0.25%) at 10 days interval
Red pumpkin beetle/borer	Spray seven 50 WP (0.4%) or Cymbush (0.05%) at the seedling stage.
Fruit fly	Bait traps should be prepared by adding to protein hydrolysate and 1ml of malathion in 1 litre of water.
Nematode	Soil application of Furodon @ 8 kg/ha33 kg/ha.

5

Bitter Gourd

B. Varalakshmi[1] and G.C. Nagesh[2]

[1]ICAR-Indian Institute of Horticultural Research, Hesaraghatta Lake Bengaluru-560089, Karnataka

[2]Namdhari Seed Private Limited, Uragahalli, Bidadi, Ramanagara, Bidadi-562109 Karnataka

Introduction-Crop Importance and Seed Market Size

Bitter gourd (*Momordica charantia* L.) is one of the important commercial cucurbitaceous cash crops grown mainly by small holding farmers in Asia where it is cultivated on more than 340,000 ha annually (McCreight *et al.*, 2013). Eastern Asia, possibly Eastern India, or Southern China is most likely the center of origin for bitter gourd (Walters and Decker-Walters 1988; Miniraj *et al.*, 1993). It is also known as bitter melon, bitter cucumber, balsam pear, bitter apple, or bitter squash (Morton, 1967). Fruits are known for abundant source of vitamins and minerals. Delicious preparations are made from fruits after stuffing and frying. It is also consumed in the form of Juice, Tea and Pills. Bitter gourd fruit abounds in phytonutrients (Dhillon *et al.*, 2017) and often used to manage type 2 diabetes as a folk medicine. Clinical studies confirmed that elevated fasting glucose levels in pre-diabetics can be reduced by supplementing the diet with bitter gourd fruits. (Amirthaveni *et al.*, 2018; Krawinkel *et al.*, 2018). Medicinal properties present in the bitter gourd fruits are used in the curing of diabetes, asthma, blood diseases and rheumatism.

The current seed market of bitter gourd in Asia values ≈ 16 million Euros. The bitter gourd seed market in India alone is 400 metric tonnes having value of 350-400 crores. 70 % area is under hybrids and 30 % is covered by open pollinated varieties.

Floral Biology and Pollination Behaviour

In bitter gourd male and female flowers are borne solitary in the leaf axils. Flowers are big, attractive (white or yellow colour petal). Flowering usually starts after 40-45 days after sowing depending on the local climatic conditions. The anthesis occurs in the morning hours.

Anthesis, dehiscence of anthers, pollen fertility and stigma receptivity is furnished in the below table

Anthesis	Dehiscence	Pollen Fertility	Stigma Receptivity
9-13.30 hr.	7-8 hr.	5-12 hr.	Starts day before anthesis & lasts up to one

Pollination Behaviour

In bitter gourd, environmental factors influence the anthesis, pollen dehiscence and fruit setting. Bitter gourd flowers in the morning to mid-day and set fruit at higher temperature of mid-day. The pollen producing capacity varies between the cultivars. Bitter gourd is cross pollinated, the extent of cross pollination varies from 60 to 80% depending on the climate and visit of the insects. It is a fact that insects do not differentiate flowers from the same or different plants at the time of voluntary pollination. Majorly bees (*Apis florea, A. dorsata, Nomioides* sp.) act as main pollinating agents. Other insects which also helps in pollination are beetles like, *Conpophilous* sp., moths like *Plasmidia* sp. etc. To achieve highest fruit set and good seed yield, placement of atleast one bee colony/hive per acre is beneficial.

Techniques of Selfing and Crossing

Most of the cucurbits are monoecious and have large flowers, making selfing and crossing is easy. The male (staminate) flowers in bitter gourd are borne solitary. The pistillate (female) flowers appear solitary in the leaf axils. One day prior to anthesis, male and female flower buds are covered separately in the butter-paper bags. Instead of covering the male flower bud with bag, often the corolla of male flower bud is tied with string or very thin wire or adhesive cellophane tape to prevent them from opening. The next day morning, the opened male flowers are collected, and pollen is dusted directly on to the stigma of the female flowers which had been bagged previously. After pollinating, the female flowers are covered with butter-paper bag and tagged with a small label on which the cross details and date of pollination are written in pencil. To obtain better fruit-set it is advised to remove the paper bag after 4-5 days of pollination. In cucurbits the fruit-set on selfing or crossing is generally poor. It is advisable to use the first or second pistillate flowers on a vine as in later stages the fruit-setting is still poorer. For increasing the fruit-set following artificial selfing or crossing, especially the latter, certain practices commonly followed are: pruning of the vine, removing other female flower buds or other set fruits from the vine on which the crossed or selfed fruit is situated and applying growth-regulators. The selfing and crossing techniques are shown in Fig 2 and Fig 3 respectively.

Breeding Objectives-Current Industry Priorities and Breeding Methods

The following are the breeding objectives and current industry priorities-

1. Development of high yielding hybrids.
2. Development of early maturing hybrids with high femaleness.
3. Development of resistant hybrids to Leaf Curl Virus (LCV) and Powdery Mildew combined with good horticultural traits.
4. Fruit quality- shiny dark green fruits with attractive tough spines and firm fruits, absence of white tip and narrow neck and good keeping quality.
5. Good plant habit- vigor and rejuvenation.
6. Development of Gynoecy based hybrids.

Breeding Methods

Various methods usually are applied in tandem to achieve breeding goals. The most general methods used for bitter gourd improvement are single-plant selection, mass selection, pedigree selection and bulk population improvement (Sirohi 1997). Pedigree selection is the common method used for the development of inbred lines with early, high yield and high-quality fruit. Heterosis breeding, backcross method and recurrent selection can also be employed to improve several traits. For selecting the desirable trait and speed up the breeding programme marker-assisted selection (MAS) can be used as a powerful tool.

Genetics of Inheritance

Traits	**Inheritance**	**Reference**
Gynoecious	Monogenic Recessive	Ram *et al.*, 2006 and Behera *et al.*, 2009
Powdery Mildew Resistance	Polygenic	Jianwen *et al.*, 2007
Fruit Color	Dominant Gene	Kole *et al.*, 2012
Fruit Spines	Single Dominant Gene	Vahab 1989
Fruit Length	Dominant Gene	Kumari *et al.*, 2015
Fruit Surface	Single Dominant Gene	Dalamu *et al.*, 2012 and Kole *et al.*, 2012
Fruit Yield	Polygenic	Kole *et al.*, 2012
Ascorbic Acid and Iron	Dominant Gene	Thangamani 2016
Fruit Fly Resistance	Dominant gene	Tewatia and Dhanakar 1996

Varietal Achievements

Major Centers of Breeding and History

The major centers of breeding bitter gourd are IARI, New Delhi, IIHR, Bangalore, IIVR, Varanasi, KAU, Thrissur, MPKV, Rahuri and TNAU, Coimbatore. These centers have developed few hybrids and varieties for cultivation.

Important Companies

BASF (25%), East-West Seed (12%), VNR (10%) and Acsen Hy Veg (3%) are having major market share for Bitter Gourd in terms of volume.

Market Segments

In India there are five major market segments based on fruit colour, fruit length and spines *viz.,* Green Long (Fruit Length of 25-30 cm), Green Medium (Fruit Length of 15-20 cm), Green Short (Fruit Length of 8-12 cm), White (Fruit Length of 20-25 cm) and Green Smooth (Fruit Length of 15-20 cm). Green Medium and Green Short are high value, Green Long and Green Smooth are medium value and white having low value in the market. Green Long occupies 30%, Green Medium 40%, Green Short 20%, Green Smooth 5% and White 5% share in the market.

Ruling Hybrids and Reasons

The following are the major hybrids grown by the farmers in different markets.

Segment	Hybrid	Key Traits
Green Long	Palee	Good yield and plant rejuvenation
	Amanshree	Early and high yield
Green Medium	US6124	Good plant rejuvenation, powdery mildew tolerant
	VNR-22	Early, high yield and good plant rejuvenation
	Pragathi	High yield, good plant rejuvenation and powdery mildew tolerant
	Rushaan	Good fruit quality, tolerant to virus and good plant rejuvenation
Green Short	VNR-28	High yield and good plant rejuvenation
	US1315	Early, high yield and good plant rejuvenation
White	US33	Good plant rejuvenation and powdery mildew tolerant
	Maya 009	High yield, good fruit quality and plant rejuvenation
Green Smooth	Raja	Good plant rejuvenation, good fruit quality and high yield

Varietal Protection

Dus Testing Guidelines

I. Subject

These test guidelines apply to all varieties, hybrids and parental lines of bitter gourd (*Momordica charantia* L.)

II. Seed Materials Required

1. The Protection of Plant Varieties and Farmers' Rights Authority (PPV&FRA) shall decide when, where and in what quantity and quality of the seed material required for testing the variety is to be delivered. Applicants submitting material from a country other than India must make sure that all customs formalities are complied with.
2. The minimum quantity of seed to be supplied by the applicant should be: Varieties, hybrids and parental lines
   - For open field cultivation: 300g or 1500 seeds (in one submission only)
3. The seed should meet the minimum requirements for germination capacity (80%), moisture content (<8%) and physical purity (98%) prescribed for certified seed in India. Especially for storage, which requires a higher standard, the applicant should state the actual germination capacity, which should be as high as possible. The seed supplied should be visibly healthy, not lacking in vigour or affected by any important pest or disease.
4. The seed material must not have undergone any treatment unless the competent authorities allow or request such treatment. If it has been treated, full details of the treatment must be given.

III. Conduct of Tests

1. The minimum duration of tests should normally be two independent but similar growing seasons with reference to the eco-system of the variety submitted for DUS test.
2. The test should normally be conducted at two different locations. If any essential characteristics of the variety cannot be observed at these places, the variety may be tested at an additional place.
3. The test should be carried out under conditions ensuring normal growth. The size of the plot should be such that plants or parts of plant may be removed for measuring and counting without prejudice to the observations which must be made up to the end of the growing period. Each test shall include 120 plants for open field cultivation, which should be divided among 3 replications. Separate plots for

observations and for measuring can only be used if they have been subjected to similar environmental conditions.

4. Test plot design

Number of rows	5
Row length	6.0m
Plant to plant distance	0.75m
Row to row distance	2.0m
Number of replications	3

5. Observations should not be recorded on plants in border rows.
6. Additional tests for special purpose may be established by the Authority.

IV. Methods and Observations

1. The characteristics described in the table of characteristics (section VII) should be used for the testing of varieties for DUS.
2. For the assessment of distinctiveness and stability, observations should be made on 30 plants or parts of plants selected randomly, which should be divided among 3 replications (10 plants in each replication).
3. For the assessment of uniformity of characteristics on the plot as a whole (visual assessment by a single observation of a group of plants or parts of plants), a population standard of 0.5% with an acceptance probability of at least 95% should be applied. In the case of a sample size of 120 plants, the number of off-types should not exceed 3.
4. For the assessment of colour characteristics, it is recommended that Royal Horticultural Society (RHS) colour chart shall be used.
5. Observations of leaf will be recorded on one leaf above the first fruit set nodes.
6. Observations on the leaf blade should be made on a fully developed leaf blade, from the 15th node upwards to 20th node.
7. All observations on the flowers should be made on flowers between the 10th and the 20th node.
8. All observations on the fruit should be made on fruits around 8-14 days after anthesis, between the 10th and 20th node.
9. All observations on the seed should be made on fully developed and dry seed after washing and drying in the shade.
10. Intensity of green colour of cotyledon should be observed just before the development of the first true leaf.

11. The bitterness of the fruit should be observed by tasting the flesh of the middle part of the fruit at marketable maturity.
12. Colour of fruit skin at ripe stage should be observed when the fruit left on the plant has turned completely yellow, orange or reddish orange.
13. Stage of recording of different observation will be as follows:

Description	Code
a. Cotyledons completely unfolded	10
b. Active vegetative phase	20
c. 50% of the flowering stage (first pistillate flower appears in 50% plant)	30
d. Fruits attaining marketable maturity	40
e. Full maturity (ripening stage)	50

I. Grouping of Varieties

1. The collection of varieties to be grown in the trial should be divided into groups to facilitate the assessment of distinctiveness. Characteristics, which are suitable for grouping purpose, are those which are known from experience not to vary, or to vary only to lesser extent, within a variety. Their various states of expression should be fairly and evenly distributed throughout the collection.
2. It is recommended that the competent authorities use the following characteristics for grouping varieties.
   a. Fruit: Length (characteristic-15)
   b. Fruit: Diameter (characteristic-16)
   c. Fruit: Color of skin (characteristic-18)
   d. Fruit: Shape in longitudinal section (characteristic-21)
   e. Fruit: Tubercles (characteristic-22)
   f. Fruit: Ridge (characteristic-24)

II. Characteristics and Symbols

1. To assess distinctiveness, uniformity and stability, the characteristics and their states as given in the Table of Characteristics should be used.
2. Notes (1-9) should be used for the purpose of recording and electronic processing of data. Each state of expression is allotted a corresponding numerical note (1-9) for the different characteristics.
3. Legend (*) Characteristics that should be used in every growing season on all varieties and shall always be included in the description of the variety, except when the states of expression of any of these characters is rendered impossible by a preceding characteristic or by the

environmental conditions of the testing region. Under such exceptional situation, adequate explanation shall be provided.

4. Type of assessment of characteristics indicated in column-7 of table of characteristics is as follows:

   MG: Measurement by a single observation of a group of plants or parts of plants.

   MS: Measurement of a number of individual plants or parts of plants

   VG: Visual assessment by a single observation of a group of plants or parts of plants.

   VS: Visual assessment by observations of individual plants or parts of plants.

V. Table of Characteristics

Sl. No.	Characteristics	States	Note	Example varieties	Stage of observation	Type of assessment
1	2	3	4	5	6	7
1.	Cotyledon: intensity of green color	Light	3	-	10	VS
		Medium (GG-137d)	5	Pusa Vishesh, Sel-5, Arka Harit		
		Dark (GG-137a)	7	Pant Karela-1, Kashi Urvashi		
2. (*)	Plant: main vine length	short viny (<2.0m)	3	Arka Harit, Pusa Vishesh, Punjab-14	50	MS
		medium viny (2 - 2.75m)	5	Pusa Do Mausami, CO-1, Phule Ujwala		
		long viny (>2.75m)	7	Preethi, HABG-22, Kalyanpur Baramasi		
3. (*)	Stem: shape	Rounded	1	-	20	VS
		Angular	2	Pusa Do Mausami, CO-1, Phule Ujwala		
4.	Stem: length of internodes of main stem (between 15th -20th node)	Short (<5cm)	3	Pusa Do Mausami, Arka Harit, Punjab -14	20	MS
		Medium (5-8cm)	5	Preethi, NDBT-7, Kashi Urvashi, Pusa Vishesh		
		Long (>8cm)	7	Hirkani, HABG-22		

5.	Stem: number of primary branches	Less (<10)	3	NDBT-7, Phule Ujwala	20	MS
		Medium (10-20)	5	Preethi, Pusa Vishesh		
		Many (>20)	7	Hirkani, Arka Harit		
6.	Leaf blade: length	Short (<6cm)	3	Pusa Vishesh	20	MS
		Medium (>6-9cm)	5	Arka Harit, Pusa Do Mausami		
		Long (>9cm)	7	Preethi, Hirkani		
7.	Leaf blade: width	Narrow (<6cm)	3	Pusa Do Mausami, Arka Harit	20	MS
		Medium (6-10cm)	5	Pusa Vishesh, NDBT-7		
		Broad (>10cm)	7	Preethi, Hirkani, Kalyanpur Baramasi		
8. (*) (+)	Leaf blade: margin	Entire	1	-	20	VS
		Serrate	3	-		
		Multifid	5	Pusa Vishesh, Preethi, Hirkani, Kalyanpur Baramasi		
9. (+)	Leaf blade: shape	Obovate	1	-	20	VS
		Cordate	2	Kashi Urvashi, Arka Harit, Pusa Vishesh, Preethi, Hirkani, Kalyanpur Baramasi		
		Oblong	3	-		
		Reniform	4	-		
10. (*) (+)	Leaf blade: number of lobes	5 lobes	3	Kashi Urvashi, Arka Harit, Pusa Vishesh, Preethi, Hirkani, Kalyanpur Baramasi	20	MS
		7 lobes	5	-		
11.	Leaf blade: depth of lobing	Shallow	3	Arka Harit, Sel-5	20	VS
		Medium	5	Pusa Vishesh, NDBT-7, HABG-22		
		Deep	7	Preethi, CO-1		

12.	Petiole: length	Short (<5cm)	3	NDBT-7, Sel-5	20	MS
		Medium (5-8cm)	5	Pusa Vishesh, Arka Harit		
		Long (>8cm)	7	Pant Karela-1, Preethi, Punjab-14		
13. (*)	Flower colour	Light yellow (YG-3a & 3b)	3	Preethi, Kalyanpur Baramasi, NDBT-9	30	VG
		Yellow (YG-7d)	5	Arka Harit, HABG-22, Pusa Do Mausami		
		Deep yellow	7	-		
14.	Ovary: length (at the day of anthesis)	Short (<1.5cm)	3	Arka Harit, Kashi Urvashi, Pusa Vishesh	30	MS
		Medium (1.5-2.5cm)	5	Pusa Do Mausami, Phule Green Gold		
		Long (>2.5cm)	7	Kalyanpur Baramasi		
15. (*)	Fruit: length	Very short (<5cm)	1	-	40	MS
		Short (5-10cm)	3	Punjab-14		
		Medium (10.1-15cm)	5	Pusa Do Mausami, Arka Harit, Pant Karela-1		
		Long (15.1-20cm)	7	Phule Green Gold		
		Extra long (>20cm)	9	Kalyanpur Baramasi		
16. (*)	Fruit: diameter	Thin (<3cm)	3	Kalyanpur Baramasi	40	MS
		Medium (3-4.5cm)	5	Phule Ujwala, Arka Harit		
		Thick (>4.5cm)	7	-		

17.	Peduncle: length	Short <5.0 cm	3	Pusa Vishesh, Preethi	40	MS
		Medium (5.0-10.0 cm)	5	NDBT-7, Meghana-2, Kashi Urvashi		
		Long(>10cm)	7	Hirkani, CO-1		
18. (*)	Fruit: colour of skin	White	1	-	40	VG
		Creamy white (142B)	2	Preethi		
		Light green (141C)	3	Arka Harit		
		Green (137A)	4	Hirkani, Pusa Do Mausami, Sel-5		
		Dark green (147C)	5	Phule Green Gold, Kalyanpur Baramasi		
		Glossy green (143C)	6	Pusa Vishesh		
19. (*) (+)	Fruit: shape of base at peduncle end	Acute	1	HABG-22, Hirkani, Phule Green Gold	40	VS
		Obtuse	2	Preethi, Arka Harit		
		Rounded	3	-		
		Flattened	4	-		
20. (*) (+)	Fruit: shape of apex at blossom end	Acute	1	Kalyanpur Baramasi, Preethi	40	VS
		Obtuse	2	Arka Harit, Pusa Vishesh		
		Rounded	3	-		
		Flattened	4	-		
21. (*) (+)	Fruit: shape in longitudinal section	Oblong	1	Hirkani	40	VS
		Ovate	2	Arka Harit		
		Spindle shaped	3	Preethi, HABG-1, Kalyanpur Baramasi		
		Club shape	4	-		
		Triangular	5	-		

22. (*) (+)	Fruit: tubercles	Absent	1	Sel-1	40	VS
		Few	3	HABG-1, Pusa Do Mausami		
		Medium	5	Kalyanpur Baramasi, Phule Ujwala, Phule Green Gold		
		Many	7	NDBT-9, Preethi, NDBT- 7		
23.	Fruit: tubercles prominence	Conspicuous	1	Pusa Do Mausami, Arka Harit, Pusa Vishesh	40	VG
		Non-conspicuous	2	NDBT-7, NDBT-9, Preethi		
24. (*) (+)	Fruit: ridge	Discontinuous	1	Preethi, NDBT-7, NDBT-9	40	VG
		Continuous	2	Pusa Do Mausami, Arka Harit, Pusa Vishesh		
25.	Fruit: bitterness	Mild	3	Pusa Do Mausami	40	VG (sensory)
		Strong	5	NDBT-9, Preethi		
26. (*)	Fruit: color of skin at ripe stage	Yellow (YG-9C)	1	Kalyanpur Baramasi	50	VG
		Orange (OG-24a)	2	Pusa Vishesh, NDBT-7, Arka Harit		
		Reddish orange (OG-N25a)	3	Kashi Urvashi		
27.	Seediness (no. of seeds/fruit)	Very less (<10)	1	-	50	MS
		Less (10-20)	3	Punjab-14		
		Medium (21-30)	5	HABG-21, Kashi Urvashi, Arka Harit		
		Many (>30)	7	HABG-22, Hirkani, Pusa Do Maushami		
28.	Seed: length	Short (<1.4cm)	3	Pusa Vishesh, NDBT-7, Sel-1	50	MS
		Long (>1.4cm)	5	Arka Harit		

29.	Seed: colour	Light brown (GY- 161A, B, C & GO-164B)	1	Arka Harit, Preethi	50	VG
		Brown (GO-164A & GO-167C)	2	Kalyanpur Baramasi, Kashi Urvashi, Punjab-14		
		Dark brown (GO- 165B)	3	HABG-22, Phule Green Gold, Kalyanpur Baramasi		
		Yellow	4	-		
		Black	5	-		
30. (+)	Seed: indentation of margin	Small	3	Pusa Vishesh	50	VS
		Medium	5	Hirkani, Phule Green Gold, Pusa Do Mausami		
		Large	7	Arka Harit, Preethi, Meghana-2		
31.	Seed surface	Smooth	3	-	50	VG
		Rough	5	HABG-1, Preethi, Phule Green Gold		

Commercial Practice of Seed Production and Seed Production Areas

Hybrid Seed Production

For exploitation of heterosis the key necessities are presence of heterotic combinations, large flower size and attractive colour, pollen producing ability of male parent and longer stigma receptivity of female parent, easy to emasculate and pollinate, attraction to insect as a means of pollinating agent, seed setting and their economic viability in production as well as adoption. Since, bitter gourd has large flower size, monoecious sex form, coloured petals (yellow/white), no emasculation is required and these help in easy pollination due to the production of sufficient pollen grains and presence of nectaries which help in supports the F_1 hybrid seed production.

F_1 Hybrid Seed Production Steps

Mainly three steps involved in F_1 hybrid seed production are:

1. Inbred line development and their production: The inbred lines are developed through exploitation inbreeding and fixing the desired traits

in them. Such developed inbred line seed is produced in isolation or by hand pollination.

2. Combining ability testing: Combining ability (gca/sca) is tested by using line × tester or diallel mating design.
3. F_1 hybrid seed production: Several techniques have been developed for hybrid seed production and it varies from crop to crop.

Hybrid Seed Production Techniques

i. **Hand Pinching and hand pollination:** This is commercially used to produce the seed of monoecy based hybrids under-open-field conditions. In this method one day prior to pollination, female flower buds are covered with butter paper during late afternoon hours in female parent and male flower buds are collected in male parent and kept overnight in wet cloth under normal room temperature. In the next day morning, before pollination, all opened male flowers are removed from the female parent and pollination is done on previously covered female flowers with the previous day collected male flower buds. The pollinated female flowers are covered with the butter paper bag and tagged with thread. The planting ratio of female and male is 3:1 for seed production. Lot of skilled & trained labour are required in pinching, pollen collection, bagging and pollination. At IARI, New Delhi, seed production of Pusa Hybrid 1 and Pusa Hybrid 2 is being done by following this method.

ii. **Hand pinching and pollination by insects:** From the female parents, male flowers are pinched off regularly prior to the day of anthesis, The male flowers from male lines are used by the honeybees and other insects (voluntary) to pollinate the female flowers of female line. The female and male parents are grown in every alternate rows. The fruit set on female parents are of hybrid and harvested for extraction of seed. The ratio of planting of female and male is 1:1, but it again depends upon the honeybee's population in plot.

iii. **Use of gynoecious sex form:** This method is commercially exploited in the private seed industries for seed production of gynoecious based hybrids. For this method crop raising under net house is compulsory. Here, the day before pollination the male flower buds are collected from male parent and kept in wet cloth overnight under room temperature. In the next day morning pollination is done on to the freshly opened female flowers in the female parent with the previous day collected male flower buds. The ideal planting ratio of female and male is 3:1. By using gynoecious lines, the pinching of male flowers, bagging and threading process can be eliminated and there will be 50% of reduction in time and labour requirements in bitter gourd hybrid seed production.

Advantages of gynoecious lines in hybrid seed production are:

a) Elimination of tedious emasculation i.e., pinching of male flowers providing correct combination of parents.
b) Providing more flexibility to breeding.
c) Facilitating quick incorporation of diverse genes for disease resistance in a hybrid.
d) Avoids the off-types problem in hybrid seed lots.
e) It also serves as a protection of one parent from stealing in the production areas.

Because of thermo-specific nature of gynoecious lines, they are unstable under high temperature and long photo periodic conditions. These homozygous gynoecious lines will be are maintained by spraying silver nitrate @ 250 ppm or sodium thiosulphate 250-500ppm to induce hermaphrodite flowers at two to four true leaf stage. F_1 hybrid seed is produced by crossing the gynoecious lines with monoecious male parent.

At ICAR-IIHR, Bengaluru Gynoecious plants have been identified and homozygous lines are being developed in different genetic backgrounds like, green/creamy white/long/medium long/short fruit types. However, ICAR-IARI, New Delhi has developed two stable gynoecious lines DBGY-201, 202 in the Pusa Do-Mousami and Pusa Vishesh background respectively. These two gynoecious lines bear only pistillate flowers and are maintained by spraying of Silver Nitrate @ 250ppm, GA_3 @1500ppm, Sodium Thiosulphate @400 ppm or through micro-propagation using MS medium. These gynoecious lines are used for the development of F_1 hybrids by raising the gynoecious (female parent) and male parent in 3:1 or 4:1 ratio and performing hand pollination.

Fig. 1. (a) Gynoecious plant bearing hermaphrodite flower after treating with 250 ppm $AgNO_3$, **(b)** Gynoecious plant showing toxicity after treating with 500 ppm $AgNO_3$, **(c)** ovary length of hermaphrodite flower after treating with 250 ppm $AgNO_3$ and untreated female flower, **(d)** male flower, **(e)** hermaphrodite flower, **(f)** female flower, **(g)** germinated pollen of hermaphrodite flower.

iv. **Sex modification using chemicals:** In bitter gourd hybrid seed can be produced by the application of chemicals for obtaining desirable sex. For F_1 hybrid seed production in bitter gourd, spray of Ethrel (2-choloro-ethyl-phosphonic acid) @ 200-300 ppm at two to four true leaf stage and another at flowering is useful in inducing the pistillate flower successively in first few nodes on the female.

The male parent is grown side by the side of female and allowed for natural cross pollination. Hand pollination is practiced in the absence of insects. For hybrid seed four to five fruit set at initial nodes are sufficient.

Isolation Distance

The Bitter gourd is a cross pollinated crop and honeybees are key pollinators, thus for maintaining purity an isolation distance all around seed field is must to separate it from fields of other varieties, fields of the same variety but not confirming the varietal purity requirement. The isolation distance of 1000 m for certified seed, 1500 m for foundation seed and at least 2000 m for breeder seed production is required. Also, an isolation distance of 1000 m for certified seed and 1500 m for foundation seed from balsam apple (Mokha): *M. balsamina* L.; Bhat Karela: (Kakrol): *M. cochinchinensis* spreng.; Jangli Karela: *M. dioica* Roxb. Ex. Willd. is required. In case pollinator and seed parent are planted in separate blocks and hand pollination is to be performed, 5m isolation distance is required between blocks of these parental lines.

Choice of Season and Areas of Seed Production

Seed crop should be raised in such a season which persists dry weather at the time of seed maturity and seed extraction. Commercial seed production is practiced in both Rainy (June-July Sowing) and *Rabi* (Sep-Oct Sowing) season by private seed companies. However, the major production happens in the rainy season. Locations also play an important role in seed production with reference to yield and quality of seed. To explore the advantage of climate, private sector seed companies are taking up their seed production in these areas *viz.*, Haveri, Koppal and Tumkur districts in Karnataka. Jalna, D. Raja and Lunar in Maharashtra and Chhattisgarh are the major seed producing areas in India.

Rouging

Seed crop is to be rouged at various stages of crop growth for removal of off-type and obviously should be carried out prior to flowering to avoid natural cross-pollination. However, fruit set and complete fruit development stages are also important. Fruit shape, colour, stripe, neck etc. are to be considered while rouging. Usually, rouging is performed at three stages *viz.*, vegetative, flowering and fruiting stage to remove the off types from the seed field. Usually, male rouging is done one week prior to pollination.

Maturity of fruit

Usually, gourds take relatively more time to attain harvestable maturity, However, bitter gourd takes relatively less time. Maturity is influenced by climatic factors and agronomic practices of crop (trailing etc.). The maturity duration is faster in summer than in rainy season.

Harvestable Maturity in Bitter Gourd

Variety	Period in days (seed to seed)
Arka Harit, Priya, Pusa Vishesh	65 days

Generally, bitter gourd takes 40-45 days to mature after pollination and 120-130 days after sowing. Other parameters like change in fruit colour also used as maturity index besides days to maturity. Upon ripening the fruit turns bright yellow colour.

Seed Extraction

Wet method: The fruit of bitter gourd is cut longitudinally and seed is scooped out. In wet method, the seed extraction from the cut fruits is done by two ways:

i) **Natural Fermentation**: The scooped seed material is kept in plastic or steel vessel at room temperature for 48 hours and stirred for 2-3 times and then seed is washed thoroughly 2-3 times with water. While seed washing is done on rough surface (mesh or sieve) for easy removal of pulp. This method is commercially used for the seed extraction of bitter gourd.

ii) **Chemical Extraction:** HCL @ 25- 30 ml or commercial H_2SO_4 @ 8-10 ml is added per 5 kg of pulp and some quantity of water is added, stirring of pulp is done to enhance the seed separation and left for 30 minutes. Upon this process, the impurities will float and seed will settle down. Seed is washed thoroughly with clean water. This method is quick but accuracy of acid and time is important.

Seed Yield

Variety, location, season and management of the seed crop influences the seed yield. The hybrid seed yield per acre is 200-250 kg.

Seed attributes of selected varieties

Variety	Seed Yield/ fruit (g)		No of seeds/ fruit (g)	1000 Seed wt.(g)
	Summer Season	Rainy Season		
Arka Harit	3-4.5	4-5	18-20	160-170
Pusa Vishesh	4.052	-	21.77	

Seed Standards

The seed standards for hybrid seed are furnished in the below table-

Factor	Seed Standards	
	Foundation	Certified
Pure seed (minimum)	98.00%	98.00%
Inert matter (maximum)	2.00%	2.00%
Other crop seeds (maximum)	NIL	NIL
Weed seeds (maximum)	NIL	NIL
*Objectionable weed seeds (maximum)	NIL	NIL
Other distinguishable varieties (maximum)	5/kg	10/kg
Germination (minimum)	80 %	80%
Moisture (maximum)	7.00%	7.00%
For vapour-proof containers (maximum)	6.00%	6.00%

Specific Requirements

Factor	Maximum permitted (%) *	
	Foundation	Certified
Off-types in seed parent	0.010	0.050
Off-types in pollinator	-	0.050
Male flowers shedding pollens in seed parent	-	0.100
**Objectionable weed plants	NIL	NIL

*Maximum permitted at and after flowering

**Objectionable weed shall be: Balsam apple (mokha): *Momordica balsamina* L.; Bhat Karela: (Kakrol): *M.cochinchinensis* spreng.; Jangli Karela: *M. dioica* Roxb. Ex. Willd.

8. Seed Processing and Machinery

After seed extraction the seeds are dried under sunlight for 2-3 days and then under normal room temperature for one week. Light weight seeds, chaffy seeds and white seeds are removed with gravity separator equipment. Then the seeds are sent to the processing plant by the growers. At the processing plant the seeds are treated with 10 ml polymer, 2g Captan or thiram and 20 ml water per kg of seed with the help of seed treatment machine and dried under normal room temperature for 3-4 hours and then packed to the required packet sizes.

Female flower bud selected a day before anthesis

Bagging of Female flower bud a day before anthesis

Covering of Female flower bud a day before anthesis

Male flower bud selected a day before anthesis

Bagging of male flower bud a day before anthesis

Covering of male flower and a day before anthesis

Fig 2. Selfing in bitter gourd

Fig 3. Crossing in bitter gourd

References

Amirthaveni, M. S. Premakumari, K. Gomathi, and R. Yang. 2018. Hypoglycemic effect of bitter gourd (Momordica charantia L.) among pre diabetics in India: A randomized placebo controlled cross over study. Indian Journal of Nutrition and Dietetics 55:44-63.

Behera, T. K., Dey, S. S., Munshi, A. D., Gaikwad, A. B., Pal., A. and Singh, I., 2009, Sex inheritance and development of gynoecious hybrids in bitter gourd (Momordica charantia L.). Scientia Horticulture 120:130-133.

Dalamu, Behera, T. K., Satyavati, T. C. and Pal, A., 2012, Generation mean analysis of yield related traits and inheritance of fruit colour and surface in bitter gourd. Indian J Hort 69:65-69.

Dhillon, N. P. S., Lin, C. C., Sun, Z., Hanson, P. M., Dolores, R. L., Habicht, S. D. and Yang, R.Y. 2017. Varietal and harvesting stage variation in the content of carotenoids, ascorbic acid and tocopherols in the fruit of bitter gourd (Momordica charantia L.). Plant Genetic Resources 15:248-259.

Jianwen, S., Xinjun, H. and Zuhua, Y., 2007, The Inheritance of Resistance to Powdery Mildew in Bitter Ground. China Vegetables, 09.

Kole, C., Olukolu, B. A., Kole, P., Rao, V. K., Bajpai, A., Backiyarani, S., Singh, J., Elanchezhian, R. and Abbott, A. G., 2012, The first genetic map and positions of major fruit trait loci of bitter melon (Momordica charantia). J Plant Sci Mol Breed 1:1. https://doi.org/10.7243/ 2050-2389-1-1.

Krawinkel, M. B., Ludwig, C., Swai, M. E., Yang, R., Chun, K. P. and Habicht, S. D., 2018, Bitter gourd reduces elevated fasting plasma glucose levels in an intervention study among prediabetics. J. Ethnopharmacology 218:1-7. http://doi.org/10.1016/j.jep.2018.01.016.

Kumari, M., Behera, T. K., Munshi, A. D. and Talukadar, A., 2015, Inheritance of fruit traits and generation mean analysis for estimation of horticultural traits in bitter gourd. Indian J Hort 72(1):43-48.

McCreight, J. D., Staub, J. E., Wehner, T. C. and Dhillon, N. P. S., 2013, Gone global: Familiar and exotic cucurbits have Asian origins. Hort Science 48:1078-1089.

Miniraj, N., Prasanna, K. P. and Peter, K. V., 1993, Bitter gourd (Momordica spp.). In: Kalloo G, Bergh BO (eds) Genetic improvement of vegetable crops. Pergamon Press, Oxford, pp 239-246.

Morton, J. F., 1967, The balsam pear-an edible, medicinal and toxic plant. Econ Bot 21:57-68.

Ram, D., Kumar, S., Singh, M., Rai, M. and Kalloo, G., 2006, Inheritance of Gynoecism in Bitter Gourd (Momordica charantia L.). Journal of Heridity 97(3):294-295.

Sirohi, P. S., 1997, Improvement in cucurbit vegetables. Indian Hort 42:64-67.

Tewatia, A. S. and Dhankar, B. S., 1996, Inheritance of resistance to melon fruit fly (Bactrocera cucurbitae) in bitter gourd (Momordica charantia L.). Indian J Agr Sci 66:617-620.

Thangamani, C., 2016, Genetic analysis in bitter gourd (Momordica charantia L.) for yield and component characters. Asian J Hort 11(2):313-318.

Walters, T. W., Decker-Walters, D. S., 1988, Balsam-pear (Momordica charantia, Cucurbitaceae). Econ Bot 42 (2):286-292.

6

Ridge Gourd

B. Varalakshmi and G.C. Nagesh

[1]*ICAR-Indian Institute of Horticultural Research, Hesaraghatta Lake Bengaluru-560089, Karnataka*

[2]*Namdhari Seeds Private Limited, Uragahalli, Bidadi, Ramanagara Bidadi-562109, Karnataka*

Introduction: Crop Importance and Seed Market Size

Ridge or ribbed gourd (*Luffa acutangula* (Roxb.)L.) is grown commercially and in homesteads for its immature fruits which are used as cooked vegetable. It is a popular vegetable both as spring-summer and rainy season crops. It is grown in tropical and subtropical countries, especially in Asia and India as an important cucurbitaceous vegetable crop (Jansen *et al.*, 1993). The immature fruit is a rich source of dietary fiber and minerals (Sheshadri, 1990). Apart from culinary properties, it has vast medicinal properties which are traditionally used for the curing of stomach ailments and fever (Burkill, 1985; Chakravarty, 1990). The genus derives its name from the byproduct 'loofah', which is used in bathing sponges, doormats, scrubber pads, mattresses, pillows and in cleaning utensils. The tender fruits are prescribed for people suffering from malaria and other seasonal fevers because it is easy to digest and act as an appetizer. Ridge gourd contains 0.5g protein, 3.4g carbohydrates, 37mg carotene and 18mg vitamin C per 100g of edible portion. Some round varieties of ridge gourd are also used for stuffing purposes.

Ridge gourd is originated from India and is grown throughout the country by many of the small and marginal farmers for their regular income. Uttar Pradesh, Madhya Pradesh, Punjab, Chhattisgarh, Bihar, West Bengal, Maharashtra and all southern states like Andhra Pradesh, Telangana, Karnataka, Kerala and Tamil Nadu are the major states in India where ridge gourd is cultivated. It is grown in an area of 1.00 lakh ha in India with a productivity of about 10-12 t/ha which is very low. The basic reasons for low productivity are; lack of quality seed of improved varieties/hybrids, lack of advanced production technology, climatic factors like high temperature and long photoperiod, diseases like downy mildew, powdery mildew and tomato leaf curl New Delhi virus, which, of late has become very serious production constraint of ridge gourd cultivation throughout the country. The seed market size for ridge gourd in India is nearly about 100 metric tonnes with a value of 80 crores.

Floral Biology, Pollination Behaviour

In ridge gourd, female flowers are borne single but male flowers appears in clusters/racemes in the leaf axils. Flowers are large, showy (yellow colour petal). Satputia cultivar of *Luffa hermaphrodita* is a hermaphrodite form which bears only bisexual flower which is rare. Flowering in ridge gourd usually starts approximately 45-50 days after sowing depending on the climatic condition. The anthesis occurs in the evening time.

Anthesis, dehiscence, pollen fertility and stigma receptivity is given below:

Name	Anthesis	Dehiscence	Pollen Fertility	Stigma Receptivity
Ridge gourd	17-20 hr.	17-20 hr.	on the day of anthesis to 2 to 3 days in cool period & 1½ day in rainy season	6 hr. before to 84 hr after anthesis

Pollination Behaviour

Anthesis, pollen dehiscence and fruit set in ridge gourd are under the influence of environmental factors. In ridge gourd anthesis takes place in the evening hours and fruit set continues till next morning to mid-day. The ability of pollen production varies among the gourds and there is good pollen production in Luffa. Ridge gourd is naturally cross pollinated and the extent of cross pollination varies from 60 to 80% depending upon the environment and the insect visit.

It is evident that insects do not differentiate flowers from the same or different plants at the time of voluntary pollination. The bees (*Apis florea, A. dorsata, Nomioides* sp.) are the main pollinating agents. The beetles like, *Conpophilous* sp., moths like *Plasmidia* sp. etc. also helps in pollination. Sometimes, even the ants help in ridge gourd pollination. One bee colony per acre is beneficial to get the maximum fruit set and seed yield.

Techniques of Selfing and Crossing

Most of the cucurbits which are monoecious and possess large sized flowers, selfing and crossing is easy. In ridge gourd male flowers occur in clusters and the female flowers are borne solitary in the leaf axils. About 24 hours prior to their opening both staminate and pistillate flower buds are covered separately in butter-paper bags. Instead of covering the male flower bud with bag, often the corolla of male flower bud is tied with string or very thin wire or adhesive cellophane tape to prevent them from opening. Sometimes in luffa, to prevent the opening of the bud a small moistened wad of cotton is placed over the tip of the corolla. The succeeding day when the buds open, the previously bagged female flowers are pollinated by dusting the pollen directly from the male flowers. After pollinating, the female flowers are covered with butter-paper bag and tagged with a small label on which the cross details and date of pollination are written with pencil. To obtain better fruit-set it is advised to remove the paper bag after 4-5 days of pollination. In cucurbits the fruit-set on selfing or crossing is generally poor. In ridge gourd, it is advisable to use the first or second pistillate flowers on a vine as in later stages the fruit-setting is still poorer. For increasing the fruit-set following artificial selfing or crossing, especially the latter, certain practices commonly followed are: Pruning of the vine, removing other female flower buds or other set fruits from the vine on which the crossed or selfed fruit is situated and applying growth-regulators. The Selfing and crossing techniques are shown in Fig 2 and Fig 3 respectively.

Breeding Objectives-Current Industry Priorities and Breeding Methods

The following are the breeding objectives and current industry priorities -

1. Development of high yielding and early maturing hybrids.
2. Development of downy mildew and powdery mildew tolerant hybrids.
3. Fruit Quality: Attractive green or dark green, tender fruits, non-bitter type with good shelf life.
4. Plant habit-good vigor and rejuvenation.
5. Development of leaf curl virus tolerant hybrids with good horticultural traits.
6. Development of hybrids with high femaleness.

Breeding Methods

Single-plant selection, mass selection, pedigree selection, bulk population improvement, heterosis breeding and back cross method are common methods used for ridge gourd improvement. Marker-assisted selection (MAS) accelerates the breeding program and is a powerful tool for selecting the desirable traits.

Varietal Achievements

Major Centers of Breeding

The major centers of breeding ridge gourd are IARI, New Delhi, IIHR, Bangalore, IIVR, Varanasi, MPKV, Rahuri and TNAU, Coimbatore. These centers have developed few hybrids and varieties for cultivation.

Important Companies

East-West Seed, BASF, VNR seeds and Advanta have a major share in ridge gourd seed market.

Market Segments

There are three major market segments in ridge gourd based on fruit length *viz.,* Green Long (Fruit Length of 35-45 cm), Green Medium (Fruit Length of 25-30 cm) and Green Short (Fruit Length of 15-20 cm).

Ruling Hybrids and Reasons

The following are the important hybrids grown by farmers in different market areas.

Segment	Hybrid	Key Traits
Green Long	Naga	Tender fruit, dark green, high yield and good plant rejuvenation
	US6001	Tender fruit, dark green and high yield
Green Medium	Anitha	High yield, Leaf curl virus tolerant and good plant rejuvenation
	Aarathi	High yield and good plant rejuvenation
	Arka Vikram	Early flowering, high frequency female line, high yield
Green Short	Rama	Early, high yield and tender fruit

Varietal Protection

DUS Testing Guidelines

I. Subject

These test guidelines shall apply to all varieties, hybrids and parental lines of Ridge gourd (*Luffa acutangula* (L.) Roxb.).

II. Materials Required

1. The Protection of Plant Varieties and Farmers' Rights Authority (PPV&FRA) shall decide when, where and in what quantity and quality of the seed material required for testing the variety, is to be delivered. Applicants submitting material from a country other than India shall make sure that all customs and phytosanitary formalities are complied with.
2. The minimum quantity of seed to be supplied by the applicant shall be:

   Varieties/ hybrids/parental lines: 250g or 1500 seeds (in one submission only)
3. The seed material shall meet the minimum germination percentage (80%), moisture content (<8%), physical purity (98%) and genetic purity (100%) as prescribed for seed certification in India. Especially for storage, which requires a higher standard, the applicant shall state the actual germination percentage, which shall be as high as possible.

4. The planting material must not have undergone any treatment unless the competent authority allow or reject such treatment. If it has been treated, full details of the treatment must be given.

III. Conduct of Tests

1. The minimum duration of tests shall normally be two independent but similar growing seasons with reference to the ecosystem of the variety submitted for DUS testing.
2. The test shall normally be conducted at two test locations. If any essential characteristic of the variety cannot be observed at these places, the variety may be tested at an additional place.
3. The test shall be carried out under conditions ensuring normal growth. The size of the plot shall be such that plants or parts of the plant may be removed for measuring and counting without prejudice to the observations which must be made up to the end of the growing period. Each test shall include a minimum of 84 plants, which should be divided among 3 replications. Separate plots for observation and for measurement, can only be used if they have been subjected to similar environmental conditions.
4. Test Plot Design

Number of rows	4
Row length	4.2 m
Plant to plant distance	60 cm
Row to Row distance	3.0 m (1.5m if grown on single trellis)
Number of replications	3

5. Observations shall not be recorded on plants in border rows.
6. Observation shall be recorded from 10 plants from each replication.

IV. Methods and Observations

1. The characteristics described in the table of characteristics (see Section VII) shall be used for the testing of varieties for DUS
2. For the assessment of distinctiveness and stability, observation shall be made on 30 plants or parts of plants, which shall be divided among 3 replications (10 plants in each replication).
3. For the assessment of Uniformity of characteristics in the plot as a whole (visual assessment by a single observation of a group of plants or parts of plants), a population standard of 1% with an acceptance probability of at least 95% should be applied. In case of a sample size of 84 plants, the number of off-types should not exceed 1.

4. For the assessment of all colour characteristics, Royal Horticultural Society (RHS) Colour Chart shall be used.
5. Observations on the leaf blade should be made on a fully developed leaf blade, from the 15th node upwards to 20th node.
6. All observations on the fruit should be made on fruits, 10-12 days after visible fruit set, between the 10th and 20th node.
7. All observations on the seed should be made on fully developed seeds collected from fully dried fruits on the plant.
8. Intensity of green colour of cotyledon should be observed just before the development of the first true leaf.
9. Stage of recording observation on specific characteristic shall be as follows:

Description	**Code**
a. Cotyledons completely unfolded	10
b. Active vegetative phase	20
c. 50% of the flowering stage	30
d. Fruits attaining marketable maturity	40
e. Fruits attaining physiological maturity	50

V. Grouping of Varieties

10. The selected varieties to be grown in the trial shall be divided into groups to facilitate the assessment of distinctness. Characteristics, which are suitable for grouping purpose, are those which are known from experience not to vary, or to vary only to lesser extent, within a variety. The states of expression (even produced at different locations) shall be fairly and evenly distributed throughout the collection.
11. It is recommended that the competent authorities use the following characteristics for grouping of ridge gourd varieties/hybrids/parental lines.

a.	Leaf	:	Leaf blade: number of lobes (characteristic 10)
b.	Fruit	:	length (characteristic 17)
c.	Fruit	:	girth (characteristic 18)
d.	Fruit	:	shape (characteristic 19)
e.	Fruit	:	skin color (characteristic 20)
f.	Fruit	:	ridge (rib) shape (characteristic 23)

VI. Characteristics and Symbols

12. To assess Distinctiveness, Uniformity and Stability, the characteristics and their states as given in the Table of Characteristics shall be used.
13. Notes (1-9) shall be used for the purpose of recording and electronic processing of data. Each state of expression is allotted a corresponding numerical note (1-9) for the different characteristics.
14. Legend

    (*) Characteristics that shall be used every growing season for the examination of all the varieties and shall always be included in the description of the variety, except when the states of expression of any of these characters is rendered impossible by a preceding characteristic or by the environmental conditions of the testing region. Under such exceptional situation, adequate explanation shall be provided.

    (+) See explanation on the Table of Characteristic in Section VIII
15. Type of assessment of characteristics indicated in column 7 of Table of Characteristics is as follows:

    MG: Measurement by a single observation on a group of plants or parts of plants

    MS: Measurement on a number of individual plant or parts of plants

    VG: Visual assessment by a single observation on a group of plants or parts of plants

    VS: Visual assessment by observation on individual plant or parts of plants.

VII. Table of Characteristics

Sl. No.	Characteristic	States	Note	Example Varieties	Stage of observation	Type of assessment
1.	Cotyledon: intensity of green color	Light	3	Jaipur Long	10	VS
		Medium	5	Arka Sujat, Arka Sumeet, Pusa Nasdar		
		Dark	7	Pusa Nutan, Phule Sucheta, Co-1		
2. (*)	Stem shape	Rounded	1	-	20	VS
		Angular	2	Arka Sujat, Arka Sumeet, Pusa Nasdar, Pusa Nutan, GARG-1, Jaipur Long, Phule Sucheta, Co-1, Deepthi		
3.	Stem pubescence	Absent	1	Arka Sujat, Arka Sumeet, Pusa Nasdar, Pusa Nutan, GARG-1, Jaipur Long, Phule Sucheta, Co-1, Deepthi	20	VS
		Present	9	-		
4. (*)	Leaf margin	Entire	1	-	20	VS
		Serrate	2	Arka Sujat, Arka Sumeet, Pusa Nasdar, Pusa Nutan, GARG-1, Jaipur Long, Phule Sucheta, Co-1, Deepthi		
5. (*) (+)	Leaf shape	Ovate	1	----	20	VS
		Orbicular	2	Arka Sujat, Arka Sumeet, Pusa Nasdar, Pusa Nutan, GARG-1, Jaipur Long, Phule Sucheta, Co-1, Deepthi		
		Reniform	3	----		
6.	Leaf: Length (Between 15- 20 nodes)	Small (<10cm)	3	Phule Sucheta	20	MS

		Medium (10.1- 15cm)	5	Arka Sujat, Arka Sumeet, Pusa Nasdar, Pusa Nutan, GARG-1, Jaipur Long, Co-1		
		Large (>15cm)	7	-		
7.	Leaf: Width (Between 15- 20 nodes)	Small (<15cm)	3	Phule Sucheta	20	MS
		Medium (15.1- 20cm)	5	Arka Sujat, Arka Sumeet, Pusa Nasdar, Pusa Nutan, GARG-1, Jaipur Long, Co-1		
		Large (>20cm)	7	-		
8	Leaf pubescence	Absent	1	-	20	VS
		Present	9	Arka Sujat, Arka Sumeet, Pusa Nasdar, Pusa Nutan		
9.	Leaf pubescence nature (Between 15- 20 nodes)	Soft	1	-	20	VS
		Hard	9	Arka Sujat, Arka Sumeet, Pusa Nasdar, Pusa Nutan, GARG-1, Jaipur Long, Phule Sucheta, Co-1, Deepthi		
10. (*)	Leaf blade: number of lobes	3 Lobes	1	-	20	VS
		5 Lobes	2	Arka Sujat, Arka Sumeet, Pusa Nasdar, Pusa Nutan, GARG-1, Jaipur Long, Phule Sucheta, Co-1, Deepthi		
		>5 Lobes	3	-		
11.	Stem: length of internodes (between 15th and 20th node)	Short (<10cm)	3	Arka Sujat, Phule Sucheta	30	MS
		Medium (10.1-15cm)	5	Arka Sumeet, Jaipur Long, Co-1		
		Long (>15cm)	7	Pusa Nutan, GARG-1		

12.	Stem: number of primary branches	Less (<10)	3	Arka Sujat, Co-1	30	MS
		Medium (10.1-15)	5	Arka Sumeet, Pusa Nasdar, Pusa Nutan, GARG-1, Phule Sucheta, Jaipur Long		
		More (>15)	7	-		
13.	Petiole length (Between 15- 20 nodes)	Short (<8cm)	3	Pusa Nasdar, Pusa Nutan	30	MS
		Medium (8.1-10cm)	5	Arka Sujat, GARG-1, Jaipur Long		
		Long (>10cm)	7	Arka Sumeet		
14. (*)	Flower colour	White	1	-	30	VG
		Light Yellow	2	Arka Sujat, Arka Sumeet, Pusa Nasdar, Pusa Nutan, GARG-1, Jaipur Long, Phule Sucheta, Co-1		
		Yellow	3	Deepthi		
15.	Ovary: length (on the day of anthesis)	Short (<4.0cm)	3	-	30	MG
		Medium (4.1- 6.0cm)	5	Arka Sujat, Pusa Nutan		
		Long (>6.0cm)	7	Arka Sumeet, Pusa Nasdar, GARG-1, Co-1		
16.	Peduncle length	Short (<8cm)	3	Pusa Nutan, Phule Sucheta, Jaipur Long	40	MS
		Medium (8.1- 10cm)	5	Arka Sumeet, Pusa Nasdar		
		Long (>10cm)	7	GARG-1, Co-1		

17. (*)	Fruit: length	Short (<20cm)	3	-	40	MS
		Medium (20.1- 30cm)	5	Arka Sujat, Pusa Nutan, Phule Sucheta, Jaipur Long		
		Long (>30cm)	7	Arka Sumeet, Co-1		
18. (*)	Fruit: girth	Small (<10cm)	3	-	40	MS
		Medium (10.1- 14cm)	5	Arka Sujat, Pusa Nutan, Phule Sucheta, Jaipur Long		
		Large (>14cm)	7	Arka Sumeet		
19. (*) (+)	Fruit: shape	Oblong	1	-	40	VS
		Elongate	2	Arka Sujat, Arka Sumeet, Pusa Nasdar, Pusa Nutan, GARG-1, Jaipur Long, Phule Sucheta, Co-1		
		Elliptical	3	Deepthi		
20. (*)	Fruit: skin color	Light Green	1	GARG-1	40	VG
		Green	2	Arka Sujat, Arka Sumeet, Co-1, Jaipur Long		
		Dark Green	3	Pusa Nutan, Deepthi		
21. (*) (+)	Fruit: shape of blossom end	Rounded	1	Arka Sumeet, Pusa Nasdar, Pusa Nutan, GARG-1, Jaipur Long, Co-1, Deepthi	40	VG
		Pointed	2	-		

22. (*) (+)	Fruit: shape of base	Rounded	1	Arka Sujat, Arka Sumeet, Pusa Nasdar, Pusa Nutan, GARG-1, Jaipur Long, Phule Sucheta, Co-1, Deepthi	40	VG
		Pointed	2	-		
23. (*)	Fruit: ridge (rib) shape	Superficial	1	GARG-1	40	VG
		Medium	2	Pusa Nasdar, Jaipur Long		
		Deep	3	Arka Sumeet, Phule Sucheta, Co-1, Deepthi		
24.	Continuity of ridge	Continuous	1	Arka Sujat, Pusa Nasdar, Pusa Nutan, GARG-1, Jaipur Long, Phule Sucheta, Co-1, Deepthi	40	VG
		Discontinuous	2	Arka Sumeet		
25. (*)	Fruit skin luster	Matt	1	Arka Sujat, Arka Sumeet, Pusa Nasdar, Pusa Nutan, GARG-1, Jaipur Long, Phule Sucheta, Co-1, Deepthi	40	VG
		Glossy	2	-		
26.	Skin softness of the marketable fruits	Mild Soft	3	Deepthi	40	VG
		Soft	5	Arka Sujat, Arka Sumeet, Pusa Nasdar, Pusa Nutan, Jaipur Long, Phule Sucheta, Co-1		
		Hard	7	GARG-1		
27. (*)	Flesh texture	Smooth	1	-	40	VG
		Soft/Spongy	2	Arka Sujat, Arka Sumeet, Pusa Nasdar, Pusa Nutan, GARG-1, Jaipur Long, Phule Sucheta, Co-1, Deepthi		
		Fibrous-Gelatinous	3	-		

28. (*)	Flesh color	White	1	-	40	VG
		Cream	2	Arka Sujat, Arka Sumeet, Pusa Nasdar, Pusa Nutan, GARG-1, Jaipur Long, Phule Sucheta, Co-1, Deepthi		
29.	Time of marketable maturity	Early (<60 Days)	3	Pusa Nutan, GARG-1	40	MS
		Medium (60.1-70 days)	5	Pusa Nasdar, Jaipur Long, Deepthi		
		Late (>70 days)	7	Arka Sujat, Arka Sumeet, Phule Sucheta, Co-1		
30. (*)	Plant growth habit	Short Viny (<3.5m)	3	Pusa Nutan	50	MS
		Medium Viny (3.51 -5.5m)	5	Arka Sumeet, Phule Sucheta, Co-1		
		Long Viny (> 5.5 m)	7	Arka Sujat		
31.	Seediness (No. of Seeds/fruit at the time of seed extraction)	Low (<100)	3	-	50	MS
		Medium (100.1-200)	5	Arka Sumeet, Pusa Nasdar, GARG-1, Phule Sucheta, Co-1, Jaipur Long		
		High (>200)	7	Arka Sujat, Pusa Nutan		

32.	Seed: Length	Small (<1.0cm)	3	-	50	MS
		Medium (1.1- 1.2 cm)	5	-		
		Large (>1.2cm)	7	Arka Sujat, Arka Sumeet, Pusa Nasdar, Pusa Nutan, GARG-1, Jaipur Long, Phule Sucheta, Co-1, Deepthi		
33.	Seed: colour of testa	Black	1	Arka Sujat, Arka Sumeet, Pusa Nasdar, Pusa Nutan, GARG-1, Jaipur Long, Phule Sucheta, Co-1	50	VS
		Gray	2	-		
34.	Seed luster	Matt	1	Arka Sujat, Arka Sumeet, Pusa Nasdar, Pusa Nutan, GARG-1, Jaipur Long, Phule Sucheta, Co-1, Deepthi	50	VG
		Glossy	2	GARG-1		
35.	100 seed weight	Low (<10g)	3	Pusa Nasdar, Pusa Nutan, Deepthi	50	VG
		Medium (10.1- 14g)	5	Arka Sumeet, GARG-1		
		High (>14g)	7	Co-1		

Commercial Practice of Seed Production and Seed Production Areas

Hybrid Seed Production

For exploitation of heterosis the key necessities are presence of heterotic combinations, large flower size and attractive colour, pollen producing ability of male parent and longer stigma receptivity of female parent, easy to emasculate and pollinate, attraction to insect as a means of pollinating agent, seed setting and their economic viability in production as well as adoption. Since, Ridge gourd is having large flower size, monoecious sex form, coloured petals (yellow), no need of emasculation, easy pollination, sufficient pollen grains and nectaries which support the F_1 hybrid seed production.

F_1 Hybrid Seed Production Steps

Mainly three steps involved in F_1 hybrid seed production are:

1. Inbred line development and their production: The inbred lines are developed through exploitation inbreeding depression and fixing the desired traits in them. Such developed inbred line seed is produced in isolation or by hand pollination.
2. Combining ability testing: Combining ability (GCA/SCA) is tested by using line × tester or diallel cross method.
3. F_1 hybrid seed production: Several techniques have been developed for hybrid seed production and it varies from crop to crop.

Hybrid Seed Production Techniques

Hand pinching and hand pollination: This method is commercially used to produce the hybrid seed in Ridge gourd. On the day of pollination in the morning (11 am to 12 pm) the male flower buds are collected from male parent and kept in wet cloth under room temperature and unopened female flower buds are covered with the butter paper bag in female parent. On the same day in evening time prior to pollination the opened male flowers are removed from female parent and pollination is done on to previously bagged female flowers of the female parent with the male flower buds collected in the morning. The pollinated flowers are covered with butter paper bag and tagged with thread. The planting ratio of female and male is 3:1 for sufficient seed production. Recently, Arka Vikram, the first public sector hybrid developed at ICAR-IIHR, Bengaluru and the seed production of this hybrid is done by this method.

Commercial F_1 hybrid seed production in ridge gourd is time consuming and cumbersome and expensive process of involving manual pinching of male flowers in female line and hand pollination. Ridge gourd hybrid seed production demands much skilled labour. By using male sterile lines, the pinching of male flowers is can be avoided. Further, when compared to fertile lines there is 50% of reduction in time and labour for ridge gourd hybrid seed production using male sterile lines.

Male sterility is of practical application in breeding of vegetables as it facilitates F_1 hybrid seed production without hand pinching of male flowers. Deshpande *et al.* (1979) from India was the first to report male sterility in ridge gourd and then by Pradeepkumar *et al.* (2007). For the first time, Pradeepkumar *et al.*, (2012) reported cytoplasmic male sterility (CMS) with two dominant restorer genes in ridge gourd. Two male sterile mutants were identified in germplasm lines of IIHRRG-12 (long fruited) and IIHRRG-28 (medium long fruited) at ICAR-IIHR (Varalakshmi and Deepak, 2017). These two *ms* sources *viz.*, IIHRRG-12MS and IIHRRG-28MS produces the rudimentary male flowers in contrast to the bright yellow flowers with fertile pollen and healthy anthers in male fertile, monoecious plants (Fig.3). After 12–16 days of emergence, these rudimentary male buds remain unopened and fall down. Using these *ms* lines the inheritance of male sterility was studied, which is cytoplasmic genic male sterility (CGMS) type, with single dominant gene either in homozygous or heterozygous condition restoring male fertility in the presence of sterile cytoplasm.

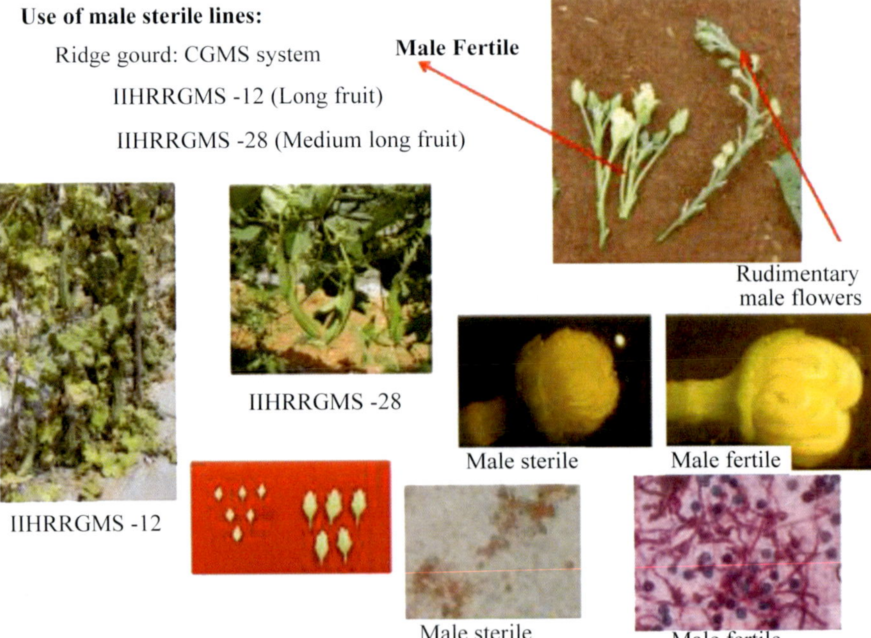

Fig. 1. Male Sterile Flowers in Ridge gourd

Cytoplasmic genic Male Sterile Line Will Solve Many Problems Like

a) Elimination of tedious emasculation i.e., pinching of male flowers providing correct combination of parents,

b) Providing more flexibility to breeding and

c) Facilitating quick incorporation of diverse genes for disease resistance in a hybrid.

At ICAR-IIHR, Bengaluru, to develop F_1 hybrids by employing male sterility, various male sterile and maintainer lines were developed in different genetic back grounds having green/dark green fruit colour and short/medium long/ long fruit length.

Isolation Distance

Ridge gourd is a cross pollinated and honeybees are key pollinator, thus for maintaining purity an isolation all around seed field is necessary to separate it from fields of other cultivars, fields of the same variety not confirming to varietal purity requirement. For hybrids, an isolation distance of 1000 m for certified seed and 1500 m for foundation seed and at least 2000 m isolation is required for breeder seed production. Also, an isolation distance of 1000m for certified seed and 1500m for foundation seed from fields of the same variety (code designation) not conforming to varietal purity requirements for certification and from sponge gourd (*L. cylindrica* Roem.) is required. Between blocks of the parental lines in case seed parent and pollinator are planted in separate blocks and hand pollination is to be adopted, 5m isolation distance is required.

For open pollinated varieties an isolation distance of 400m for certified seed and 800m for foundation seed and at least 1000m isolation is required for breeder seed production.

Choice of Season and Areas of Seed Production

Seed crop should be raised in such a season which persists dry weather at the time of seed maturity and seed extraction. Commercial seed productions is practiced in both Rainy (June-July Sowing) and *Rabi* (Sep-Oct Sowing) season by private seed companies. However, the major production happens in the rainy season. Locations also play an important role in seed production with reference to yield and quality of seed. To explore the advantages of climate, private sector seed companies are taking up their seed production in these areas. Haveri and Koppal in Karnataka and Jalna, D. Raja and Lunar in Maharashtra are the major seed producing areas in India.

Rouging

Seed crop is to be rouged at various stages of crop growth for removal of off-type and obviously should be carried out prior to flowering to avoid natural cross-pollination. However, fruit set and complete fruit development stages are also important. Fruit shape, colour, stripe, neck etc. are to be considered while rouging. Usually, rouging is performed at three stages *Viz.*, vegetative, flowering and fruiting stage to remove the off types from the seed field. Usually, male rouging is done one week prior to pollination.

Maturity of Fruit

Gourds take relatively longer time to attain harvestable maturity. Maturity is influenced by climatic factors and agronomic practices of crop (trailing etc.). The maturity duration is faster in summer than in rainy season. The ridge gourd takes 70-80 days after pollination and 140-150 days from the date of sowing to maturity. The fruit should dry completely and turn gray colour after maturity. When we shake the dried fruit sound will come.

Harvestable Maturity in Ridge Gourd

Variety	Period in days (seed to seed)
Arka Prasan, Arka Vikram, Arka Sujath, Arka Sumeet	135-150 days

Seed Extraction

Dry method of seed extraction is followed and this method is commercially used for seed extraction in Ridge gourd. From the blossom end side, the fruits are cut manually and the seeds will come out from the fruit at once.

Seed Yield

The hybrid seed yield per acre is 280-300 kg.

Seed yield varies upon the variety, location, season and management of the crop.

Seed Attributes of Selected Varieties

Crop	Variety	Seed Yield/ plant (g)		No of seeds/ fruit	1000 Seed wt.(g)
		Summer Season	Rainy Season		
Ridge gourd	Arka Sumeet	60-70	50-60	100-150	160-170
	Arka Prasan	40-50	50-60	60-90	120-140

Seed Standards

The seed standards for hybrid seed are furnished in the below table

Factor	Seed Standards	
	Foundation	Certified
Pure seed (minimum)	98.00%	98.00%
Inert matter (maximum)	2.00%	2.00%
Other crop seeds (maximum)	Nill	Nill
Weed seeds (maximum)	Nill	Nill
*Objectionable weed seeds (maximum)	Nill	Nill
Other distinguishable varieties (maximum)	5/kg	10/kg
Germination (minimum)	80 %	80%
Moisture (maximum)	7.00%	7.00%
For vapour-proof containers (maximum)	6.00%	6.00%

Specific Requirements

Factor	Maximum permitted (%) *	
	Foundation	Certified
Off-types in seed parent	0.010	0.050
Off-types in pollinator	-	0.050
Male flowers shedding pollens in seed parent	-	0.100
**Objectionable weed plants	Nill	Nill

*Maximum permitted at and after flowering

Seed Processing and Machinery

After seed extraction the seeds are dried under sunlight for 2-3 days and then under normal room temperature for one week. Light weight seeds, chaffy seeds and white seeds are removed with gravity separator equipment. Then the seeds are sent to the processing plant by the growers. At the processing plant the seeds are treated with 10 ml polymer, 2g Captan or thiram and 20 ml water per kg of seed with the help of seed treatment machine and dried under normal room temperature for 3-4 hours and then packed to the required packet sizes.

Female flower bud selected on the day of anthesis

Bagging of Female flower bud on the day of anthesis

Covering of Female flower bud on the day of anthesis

Male flower bud cluster selected on the day of anthesis

Bagging of male flower bud cluster on the day of anthesis

Covering of male flower bud cluster on the day of anthesis

Fig 2. Selfing in ridge gourd

Male and Female flower on the day of anthesis

Removal of petals of Female flower

Depetalled female flower

Tagged selfed/crossed female flower

Removal of petals of male flower on the day of anthesis

Pollination on the day of anthesis in the evening

Bagging of selfed/crossed female flower

Fig 3. Crossing in ridge gourd

References

Burkill, H.M. 1985. The Useful Plants of West Tropical Africa, Second ed. Royal Botanical Gardens, Kew. The Whitefriars Press Limited, London.

Chakravarty, H.H. 1990. Cucurbits in India and their role in development of vegetable crops. In: Biology and Utilization of the Cucurbitaceae. Bates, D.M., Robinson, R.W., Jeffrey, C. (Eds.) Cornell University Press, Ithaca, New York, pp.325-348.

Deshpande, A.A., H. Ravishankar and Bankapur, V.M. 1979. A male sterility mutant in ridge gourd (*Luffa acutangular* Roxb.). Curr. Res. : 97–98.

Jansen, G. J., Gildemacher, B. H. and Phuphathanaphong, L., 1993. *Luffa* P. Miller. In: Siemonsma, J. S. and Kasem Piluek (Editors). Plant Resources of South-East Asia No 8. Vegetables. Pudoc Scientific Publishers, Wageningen, Netherlands. pp. 194-197.

Pradeepkumar, T., Sujatha, R., Krishnaprasad B.T. and Johnkutty, I. 2007. New Source of male sterility in ridge gourd (*Luffa acutangula* (L.) Roxb.) and its maintenance through in vitro culture. Cucurbit. Genet. Coop. Rpt. 30: 60-63.

Pradeepkumar T., Hegade, V.C., Kannan, D., Sujatha, R., George, T.E. and Nirmaladevi, S. 2012. Inheritance of male sterility and presence of dominant fertility restorer gene in ridge gourd (*Luffa acutangula* (Roxb.) L.). Scientia Horticulturae. 144: 60-64.

Sheshadri, V.S. 1990. Cucurbits. *In*: Vegetable Crops in India. Bose, T.K., Som, M.G. (Eds.), Nayaprakash publishers, Calcutta, pp. 91–164.

Varalakshmi B and G. C. Deepak. 2017. Identification of male sterility, its inheritance and fertility restoration in Ridge gourd [*Luffa acutangula* (Roxb.) L.]. Abstracts of International Symposium on Horticulture: Priorities and Emerging Trends, 5-8th September, 2017, Bengaluru, India.pp.182.

7

Pumpkin

Raja Shankar S

Division of Vegetable Crops, ICAR-Indian Institute of Horticultural Research Bengaluru, Karnataka

Introduction

Pumpkin ($2n = 40$) is an important 'pepo' vegetable of the Cucurbitaceae family. Five domesticated species *viz.*, *C. maxima*, *C. moschata*, *C. pepo*, *C. argyrosperma* and *C. ficifolia* are included in the *Cucurbita* genera (Schaffer and Paris, 2003). Owing to the presence of maximum genetic diversity in the neotropical regions and its ability to grow in varied agro-ecological conditions and its tolerance to hotter climatic conditions as compared to the other cultivated *Cucurbita* species, *C. moschata* is known as "tropical pumpkin" (Andres 2004). In terms of the average nutritive values, pumpkin (2.8) stands at a higher position when compared to many widely grown vegetables. The fruits are rich in carotenoid pigments. However, the proportion of beta carotene is higher than the others. Beta carotene shows strong anti-oxidative and anti-cancer properties. While, it has relatively low content of other carotenoids like alpha-carotene, lutein, zeaxantin and violaxantin (Matus *et al.*, 1993; Fraser and Bramley, 2004). Moreover, pumpkin fruits are also rich in mineral elements *viz*., potassium, phosphorus, magnesium, iron and selenium. The mature pumpkin seeds contain the essential unsaturated fatty acids and minerals like zinc. Pumpkin-rich diet has also shown to have a pharmacological effect on reducing blood glucose (Xiong and Cao, 2001; Cai *et al.* 2003). Further, water-extracted pumpkin polysaccharides showed better hypoglycemic (Zhang and Yao, 2002). As a result, pumpkin has emerged as a staple food in many developing countries (Ferriol and Pico, 2008).

Pumpkin is an important tropical crop grown almost throughout the year in different parts of India. It is grown an area of 0.71 million hectare and produced 171 million tonnes with productivity of 21.97 tonnes/hectare (2017-18) which is lower than that of countries like Indonesia (68.3 t/ha) and the Netherlands (51.06t/ha). The higher productivity in these countries is probably due to the coverage of maximum area by the hybrid varieties against the predominance of open-pollinated varieties in India.

Taxonomy

Cucurbita sp. has chromosome number of 20 pairs (2n=40) invariably in all species. Before the discovery of America, the present day cultivated species *viz*., *C. pepo*, *C. moschata*, *C. mixta, C. maxima* and *C. ficifolia* were selected by American Indians (Whitaker and Bemis 1976). Among these, *C. moschata* seems to be the possible common ancestor of the members of the genus *Cucurbita* from which the other four cultivated species evolved independently as separate entities. Phylogenetic analysis indicated *C. moschata* and *C. mixta* as the probable sister species. Further, *C. pepo* shares a common ancestor with these species that is not shared with *C. maxima* (Decker-Walters *et al.* 1990).

Botanical Characters of Pumpkin, Summer Squash and Winter Squash

Characters	*C. moschata*	*C. pepo*	*C. maxima*
Growth habit	Annual vine	Annual vine or Bush	Annual vine
Sex expression	Monoecious	Monoecious	Monoecious
Leaf texture	Soft, hairy	harsh and prickly	Non-harsh and prickly
Leaf shape	5-6 shallowly lobed leaves with white patches on the supper surface	Broad, obtuse or acute, usually with 5 deep lobes and dentate margin. The upper surface is glabrous and lower surface has glandular club shaped hairs	More or less reniform with 5 rounded, shallow lobes with minutely dented margins
Tendril	Tendril many fid	Branched	Tendril 2-6 fid
Flower	Solitary	Solitary	Solitary
Peduncle	Smooth, 5 angled expanded or flared at fruit attachment	5 angled with little or no expansion at the fruit attachment	Spongy, cylindrical, soft, corky
Fruit	Usually large, variable shape and size (globular, cylindrical, flattened)	Usually of variable size, shape and colour	Usually large, oblong-cylindrical or flattened-cylindrical
Seed	Yellow with a thin or ragged margin, scalloped or shredded in appearance, the margin more deeply coloured than the body of the seed	Whitish yellow, broadly or narrowing ovate usually with raised, smooth, well-differentiated margin	White or yellowish with margins of different colour

Floral Biology

Pumpkin is a monoecious crop where a succession of development of staminate and pistillate flowers are observed as the plant matures. Flowers are unisexual, solitary, axillary large and showy. The fruit is a hard-rind berry, botanically called 'pepo'. The pericarp with a very little portion of the mesocarp is generally consumed. The flower stalk of staminate flowers is longer and thinner than that of the pistillate flowers. In general, across pumpkin genotypes, staminate flowers are borne before the pistillate flowers with few exceptions. The number of staminate flowers to pistillate flowers varies across genotypes ranging from 4:1 to 20:1. Accordingly, the anthesis of staminate flowers is earlier (55-75 daysafter sowing) than the pistillate flowers (60-75 days after sowing).

Breeding Methods

The primary breeding objective of pumpkin is to obtain higher harvest index per unit area, development of short vines/ bush type with predominance of pistillate sex expression, higher fruit yield, earliness as indicated by the appearance of the first pistillate flower at lower node, fruits with high β-carotene, thick flesh and sugar contents coupled with resistance to important diseases and insect pests. Considering the industrial importance of this crop, the seed yield; seed size and oil content with rich nutrients are also prioritized as major breeding objectives.

Individual plant selection: Pumpkin is a cross-pollinated crop, however, due to inbreeding it generally do not show significant loss in vigour. Hence, inbreeding followed by individual plant selection through pedigree method can be practiced as effectively as is followed in self-pollinated vegetable crops. Accordingly, the heterozygous progeny of the open-pollinated cultivars segregates in successive generations and individual plant selection was used for improvement of different traits. Damarany (1989), after following three generations of inbreeding, recorded such improvement for fruit weight, yield and resistance to powdery mildew disease. Marked improvement for fruit yield per plant (by 89.4%), number of marketable fruit per plant (by 85.9%) carotenoids content (by54.2%) was recorded after four generation of self-pollination over the base population (Mandal, 2006). Further, moderate improvement (ranging between 7.8 to 22.7%) was also recorded for traits *viz.*, polar and equatorial diameter fruits, fruit weight, hundred seed weight and total sugar content of the pulp. In earlier experiments, Kubicki and Walczak (1976) also recorded the improvement in β-carotene content by 70% by self pollinating the cultivars of *C. maxima* followed by selection for higher β-carotene content in the flesh. From another experiment, a Gibberellic Acid related dwarf mutant 'Non-vine I' was identified (Cao *et al.* 2005). In this mutant, the dwarfness is the resultant of the failure of elongation process of the cells of the internode.

Combination breeding: Desirable parental pumpkin lines are hybridized followed by self-pollination and selection for superior plants are also followed to ensure recombination of genes from both the parents. In absence of morphological manifestation of gene recombination, back cross breeding is attempted to transfer one or more desirable traits or genes from the donors into desirable and elite genetic backgrounds. The backcrossing s generally followed by few generations of self-pollination to attain uniformity of traits, after which, the genetically uniform breeding lines are self-pollinated or sib-pollinated and seeds from many selected plants are bulked (Whitaker and Robinson 1986). In pumpkin, for identification of desirable line by evaluating a sufficiently large segregating population is not generally followed as this requires large area due to the requirement of wide row to row and plant to plant spacing.

Heterosis breeding: Heterosis is the phenomenon where the hybrids resulting from the crossing between the distant parents are superior over and above the average their straight-bred parents. Hence, F_1 hybrids generally have increased vigor, improvement in fruit yield, uniform and early maturity over the over the open pollinated varieties. The effectiveness of hybrid is determined by the genetic purity of the parental lines and heritability of the characters under selection. In this method, the parents are selected based on their specific Combining Ability (SCA) as the General Combining Ability (GCA) is the characteristic of parents and SCA is characteristic of crosses or hybrids. When compared to the open-pollinated varieties, hybrids generally show 15 25% improvement in yield (Virmani *et al.*, 2006).

The superior performance of hybrid over its parents are estimated either based on a mid-parent value which is known as Mid Parental Heterosis (Falconer and Mackey, 1996) or better parent value which is known as Better Parental Heterosis (Fonseca and Patterson, 1968) or superior standard check variety of the region which is known as Standard Heterosis. For generation of the experimental hybrids, line × tester or full- or half-diallel design is conventionally followed to (Durrudi *et al.*, 2018). Generally, the GCA provides insights into the additive genetic variance, and the SCA provides insights into the non-additive effects such as dominance and epistasis. Thus, in heterosis breeding programmes, a breeder emphasizes on the selection of a cross combination with high SCA among superior parental lines with high GCA (Virmani, 1994). Earlier, many investigators have found significant heterosis for earliness, number of fruits per plant, average fruit weight, fruit flesh thickness and fruit yield in pumpkin (Nisha *et al.*, 2014; Tamil Selvi and Jansir Rani, 2016; Jansi, 2018; Kumar *et al.*, 2018; Darrudi *et al.*, 2018; Po-Ya-Wu *et al.*, 2019; Hosen *et al.*, 2022; Shafin *et al.*, 2022).

Heterosis for yield related traits: Tamilselvi *et al.* (2014) used twelve lines and three testers in a line × tester mating design to generate experimental hybrids and compared their performance with that of the commercial check MPH-1. Positive and significant heterosis was observed for the number of fruits per vine, thickness of flesh, seed yield per fruit and fruit yield per vine in the hybrid resulting from Kasi Harit × Avinashi Local followed by the hybrid resulting from Vadhala Gundu Local × CO-2 which also showed positive heterosis for the above-mentioned traits and carotenoids content. In another study, Singh *et al.* (2018) identified the genotype P-3621 as the best general combiner shorter vine and internode, equatorial diameter and high fruit shape index while the genotype P-6242 was best for pulp thickness, P-2211 was best with respect to the number of primary branches per plant, P-10224 was best for polar diameter and vegetative growth and P-1343 was best for petiole length. However, preponderance of high SCA in the crosses *viz*., P-10224 × P-6242, PS × P-6242, P-41212 × P-2211, PS × P-364, P-41212 × P-6242, P-10224 × P-2211 indicated the predominance of both additive and non-additive gene actions in the inheritance of bushy and butternut traits. Thus, these traits can be effectively improved by heterosis breeding as well as recurrent selection. Further, there were three other hybrids with higher pulp thickness and on par yield potential with the standard check. Negative heterosis is important in improvement of earliness traits. Marxmathi *et al.* (2018) observed negative and significant standard heterosis for earliness in CM-23×CM28. Further, the cross CM1×CM23 showed high positive significant heterosis for fruit length, average fruit weight and for number of seeds per fruit while the cross CM29×CM1 showed significant negative heterosis for fruit length. The heterosis was positive and high for the fruit diameter in CM1×CM28 while it was negative and significant in CM29×CM28. The cross CM28×CM31 showed significant reduction in the number of seeds per fruit. The maximum positive and significant heterosis for fruit yield was recorded in CM23×CM1 followed by CM12×CM28. Stability and homozygosity of the parental inbreds also have a profound effect on the performance of the resultant hybrids. In this direction, Restrepo *et al.* (2018) found the hybrids resulting from the inbred lines which were self-pollinated twice were superior and heterotic than the inbreds which were self-pollinated once. The highest expression of heterosis for yield per plant was observed in hybrid resulting from UNAPAL-Abanico-75-1 × UNAPAL-Dorado followed by UNAPAL-Dorado × UNAPAL-Llanogrande-2 and UNAPAL-Abanico-75-1 × UNAPAL-Abanico-75-2. Moshin *et al.* (2022) employed full diallel mating design including six diverse pumpkin parents and generated 30 experimental hybrids from which two hybrids *viz*., CM-1×BARI Mistikumra-1 and CM-8 × CM-1 were found most promising which showed significantly higher heterosis, heterobeltiosis and standard heterosis. Based on these findings, cross combinations involving indigenous and exotic lines might be desirable for improvement of fruit yield through the development of commercial hybrids.

Heterosis for quality traits: Shafin et al (2022) found IBD40×IBD47 as the best specific combiners for beta carotene content, fruit yield and total sugar content; IBD23×IBD40 was best for TSS, flesh thickness and hollowness; IBD40×IBD57was best for fruit breadth; IBD47×IBD50 was best for non-reducing sugar content and IBD47×IBD57 was best for reducing sugar content. In a similar study, Hosen *et al.* (2022) found Gold Butter 315 to possess maximum general combining ability for dry matter content and antioxidant properties; Asian pumpkin for total carotenoids; Sarawak for the number of fruits number per plant and Australia-1 for flesh thickness, average fruit weight, fruit yield per plant and TSS. Maxmathi *et al.* (2018) recorded high heterosis for TSS, content of beta-carotene and dry matter was found in cross combinations *viz.*, CM1×CM31, CM29×CM28, CM28×CM29 and CM28XCM23.

Varieties and hybrids of pumpkin: There are several cultivars have been developed in the country and grown. The few popular cultivars are Arka Chandan (pumpkin) and Arka Suryamukhi (winter squash), Pusa Vishwas, Pusa Vikas, Kashi Harit, CM-14, CO-1, CO-2. In addition, the hybrids such as Pusa Hybrid-1 (pumpkin) and Pusa Alankar (summer squash) are also most popular in the country.

Hybrids available in the market: There are several commercial hybrids available in the market recommended by private companies for different traits.

F_1 Hybrids	Fruit shape/colour	Fruit weight (kg)	Source
Karna-F_1	Round fruit with patches	4-5kg	Sagar Biotech Pvt. Ltd.
Ashoka- F_1	Round fruit with no patches	4-5kg	Sagar Biotech Pvt. Ltd.
Mittra-920	Globular with patches	4-5kg	Puregene Agritech Pvt. Ltd.
VNR-P-11	Globular with patches	6-8kg	Sagar Seeds Pvt. Ltd.
Sarpan-901	Globular with patches	7-9kg	Sarpan Seeds
Urja Amrit	Flat round with patches	3-4kg	Urja Seeds
Urja F_1-64	Globular with patches	4-6kg	Urja Seeds
Jivit small	Globular with patches	4-6kg	Jivit Seeds
Jivit Large	Oval with patches	10-12	Jivit Seeds
Iris-9060	Oblong with whitish patches	5-7kg	Iris hybrid seeds
Iris Beema F_1	Flat round with dark green	4-6kg	Iris hybrid seeds
Iris-716	Flat round with dark green and warty skin	4-5kg	Iris hybrid seeds

Nirvana F_1	Round with dark green	4-6kg	Shine brand seeds
Remik God	Flat round with yellow stains	2-2.5kg	Remik Seeds
Ganga Phal	Flat round with patches	3-4kg	Riccia seeds
SVHP-077	Flat round with patches	4-5kg	Sakthi Vardak hybrid seeds Pvt. Ltd.
SVHP-088	Oblong with patches	5-6kg	Sakthi Vardak hybrid seeds Pvt. Ltd.
MPH-1	Oblong with patches	4-6kg	Mahyco Hybrid Seeds
Anuj F_1	Flat round with patches	4-5kg	VNR Seeds

Commercial Hybrid Seed Production Method in Pumpkin

The demand of hybrid seed among farmers is increasing due to the attributes like earliness, uniformity and higher yield, which make pumpkin cultivation more profitable than traditional/open pollinated varieties. Commercially, hybrid seeds of pumpkin are mostly produced by the following methods:

Protection of Both Staminate and Pistillate Flowers Coupled with Hand Pollination

The pistillate flower of the seed parent and staminate flower of the pollen parent are protected from opening by covering them with butter paper bag in the previous day evening or by tying the corolla tube to protect the flowers from foraging insects. In the following morning, pollens are collected from the of staminate flowers of male parent and are transferred to the stigma of the pistillate flowers of the seed parent by gently touching the anthers on the stigma without causing injury. Hand pollination between 5.00 am to 7.00 am is most effective than late morning as the stigma receptivity is peak between these times. The success in fruit set upon manual cross pollination ranges widely between 65.3 to 82.5% with the average of 73.9%. The number of hybrid seeds per successful cross pollination differs widely from 124.3 to 356.5 with mean value of 256.8 per fruit. The plants of the seed parent allowed to grow only two shoots per plant without any more secondary branches. Such plants are allowed to produce 3-4 pistillate flowers per plant for hand pollination. Pinching off terminal shoot encourages seed development in hand-pollinated flowers and greater seed yield (Kalloo, 1988).

Manual Defloration of Staminate Flowers in Female Parent with Open Pollination with Male Parent in Isolated Field

If the hybrid seed production plot is properly isolated by time and space, one row of the selected pollen parent in interplanted with three rows of the seed parent. The staminate flower buds are removed from the seed parent and special care is taken to remove the staminate flower buds regularly from the seed parent until there are 2-3 fruit set in every plant of the seed parent by cross pollination through honey bees. After such fruits are set, the production of subsequent flowers can be stopped by cutting the apical growing point of the vines. Keeping of bee hives are recommended in such commercial seed production plots to maximize seed yield.

Chemical Suppression of Staminate Flowers and Open Pollination

To maximize the production of pistillate flowers in seed parent, the sex expression of monoecious pumpkin can be altered by the application of plant growth regulators exogenously. Ethrel, an ethylene-releasing chemical can be sprayed @ 250 ppm at two and four true leaf stages enhances the production of pistillate flowers. Similarly, growth regulaotors can also be used for the prevention of appearance of staminate flowers for a longer period in the vine of the intended seed parent to ensure reduction in cost of labour in hybrid seed production. Gerdakaneh *et al.* (2018) found that spraying NAA and paclobutrazol at two and four true leaf stage increased the production of pistillate flowers, the number of fruits per plant, number of seeds per fruit, test weight and total seed yield per hectare. However, there was a reduction in the number of staminate flowers and ratio of staminate to pistillate flowers when compared to the non-sprayed plants. Spray of paclobutrazol @ 50 mg/l resulted in the highest yield of both fresh and dried seeds of pumpkin.

Utilization of Male Sterile Lines

Utilization of male sterile lines as seed parent can efficiently prevent self-pollination in commercial hybrid seed production. Thus, utilization of male sterility economizes hybrid seed production. Genetic Male Sterility eliminates the requirement hand emasculation or defloration of the seed parent in hybrid seed production plots. However, the identification of 50% male fertile plants with fertile staminate flowers becomes crucial as the Genetic Male Sterile lines segregates into 50% fertile and 50% sterile plants. The highest percent of male sterile plant that can be expected is 50%, produced by crossing plants heterozygous (ms/+) with those homozygous (ms/ms) for the male sterile gene. Two single recessive genes governing male sterility in *C. pepo* have been reported (Eisa and Munger, 1968). Male sterility has also been reported *C. maxima* (Scott and Riner, 1946). Utilization of Cytoplasmic Male Sterility eleimnates such problems as the complete progeny resulting from the male sterile and male fertile plants become male sterile in this system. However, till date, the source of has not been reported in the genus *Cucurbita*. Bocharnikov *et al.* (2017) created new maternal lines *viz*., CL-fms (with cut-leaves and choriphyllous and EL-fms (with entire-leaves and integrifolious) with general combining ability and functional male sterility making these lines are suitable for the production of early and large-fruited F_1 hybrids of pumpkin. As these developed lines were functionally sterile, the pollens collected from the forced-open anthers of the flowers showed a high fertility of 90.2%. Thus, such pollens can be used for manual pollinations for self-pollination of the line for its maintenance. Further genetic analysis demonstrated the inheritance of functional male sterility in the large fruited pumpkin through a single recessive gene. Further, a high General Combining Ability was observed for the ripeness trait in the CL-fms and EL-fms lines.

Agro-techniques for Hybrid Seed Production

Soil type: It requires fertile, well aerated soil with a pH ranging between 5.5 and 6.5 for the best growth. For superior growth and yield, the soil should contain high amount of organic matter. Well-drained sandy-loam soil favours good vegetative growth and subsequent fruit development as compared to the growth and development in the clay soil in which poor plant growth is evidenced. Ploughing of the field is done for 3-4 times to obtain a fine tilth of soil. The planting can be done either in flat bed or ridge-furrow system at a recommended spacing.

Isolation distance: The sex expression of pumpkin is monoecious. Thus, the staminate and pistillate flowers are borne on the different nodes of the same plant and are cross pollinated through the insects, predominantly by the honey bees. Raising of parental lines in isolation (1000 m) from other variety is recommended to maintain genetic purity. In hybrid seed production plots for hand pollination, raising the seed and pollen parental lines in separate blocks at least 5 m distances from each other is advocated. Field standard for seed production of varieties of pumpkin and squash suggests 1000 m isolation distance for production of foundation seed and for 500 m for production of certified seed.

Climate: Pumpkin grows best in sunny areas with frost-free growing period of 4-5 months. The ideal temperature ranges from 28-32° C, where high temperatures (above 35° C) cause reduction in pollination and fruit set and low temperatures below 15° C delay the seed germination (happens 14 days) and vegetative growth. High humidity with low temperature favors more fungal pathogenic infestation. Frost cause serious injury to the plants, hence sowing of seeds in the winter season should be avoided.

Seed rate and spacing: In hybrid seed production, uniform plant stands both in the blocks of seed and pollen parent are essential to ensure maximum seed yield. About 0.5 kg seeds of pollen parent and 1.50 kg of seed parent are sufficient for one hectare of land. Before sowing, the seeds should be treated with fungicides like captan or thiram @ 2.5 gram per kg of seeds. The row to row spacing should be kept 4.0 meters and the plant to plant spacing should be kept 1 meter for ease in different intercultural operations like weeding, hoeing as well as for pollination and inspection. To maintain uniform crop stand, sowing of more than one seeds are recommended. However, after germination, the extra seedlings should be removed from the hills and only one seedling per hill should be kept.

Pumpkin is planted at different spacing while *C. pepo* is planted at lesser spacing as compared to *C. moschata* and *C. maxima*. Generally, 2.5×2.0 m spacing is optimum. For *C. pepo* 2.5×1.0 m is ideal.

Propagation: Seeds from freshly harvested fruits show higher and uniform seed germination in the main field when compared to the old seeds. However, to ensure more uniform germination, soaking of seeds in water for at least eight hours is recommended. Sowing of the seeds can be done directly in the main field or the seedlings may be first raised in the propagation trays or plastic bags and the seedlings can be transplanted after two weeks of sowing. In direct sowing, pits of 45cm^3 are prepared at recommended spacing and filled with a mixture of farmyard manure 10 kg. Two to three seeds should be sown in each pit at a depth of 2.0 cm with sufficient spacing within the pit. While sowing, care must be taken to place the pointed side of the seed facing down as it hastens germination. Under ideal conditions, the seeds start germinating after 4 four days of sowing. After germination, the most healthy and vigourously growing seedling is maintained in each hill and the unhealthy and poorly growing seedlings are thinned out. For propagation tray method, the trays having hole size of 4.0 × 4.5 cm are optimum to raise the pumpkin seedlings. To avoid the occurrence of various soil borne pathogens, the potting media intended to be used should be fully sterilized before filling the propagation trays. Until transplanting, optimum moisture should be maintained in the plugs by providing light irrigation at regular intervals. Both under irrigation and over irrigation are harmful where the under irrigation may result in poor growth of the seedlings while overirrigation may provoke occurrence of diseases like root rot and damping off.

Method of planting: In direct sown method, the planting of seed and pollen parent needs proper care without mixing them together in the main field. To avoid such chances, different labours should be engaged for sowing of seed and pollen parent. In general, one of the two available methods *viz.*, single and double block methods of sowing are followed for hybrid seed production. In the single block system, the seed parent and pollen parent are sown in 3: 1 ratio within the channels in the repeated manner in the same block. However, for easy identification, the hills sown with the pollen parent are marked with the painted wooden sticks.

In double block system, the seeds of the seed parent and pollen parent are planted in 75: 25 proportions in separate blocks. However, the male block is marked and collection of pollen donors is made from the pollen parent block at the time of pollination. Thus, this method helps to ensure higher genetic purity of the produced hybrid seed lot by avoiding physical contamination and chances of mechanical mixture at different stages of production, harvesting as well as post-harvest handling.

Fertilizer management: No standard fertilizer recommendation is followed in pumpkin. Pumpkin respond best for farmyard manures and results vigorous growth. Hence, about 1.0 kg well rotten farmyard manure, 50 g DAP and 25 g MOP per hill should be applied as basal dose at about one week before planting/sowing of seeds. As top dress, 50g urea per hill should be mixed around the plant at 30-35 days after sowing. A general fertilizer dose of 100:100:40 kg NPK/ha is recommended. After commencement of flowering in the seed parent, application of heavy dosage of nitrogenous fertilizers should be avoided to minimize the production of staminate flowers and check the excessive vegetative growth. To maximize the fruit set and seed development, foliar spray of 1% urea or DAP is also recommended during the fruit development stage.

Irrigation management: Pumpkin is classified as a deep-rooted crop of which the roots may penetrate even up 1.5 m. Being a deep-rooted plant, it can tolerate soil moisture stress to certain extent. However, to maximize the seed yield, irrigation at regular interval is recommended throughout the growing season. To enhance the efficiency of irrigation water, drip irrigation may also be followed in paired row planting. For irrigation, the turgidity of the leaves may be considered as an indicator of moisture requirement. Like most other fruit vegetable crops, the most critical stage of irrigation in pumpkin is flowering and fruit development stage. Thus, it is recommended to provide sufficient moisture this stage. Moreover, it is also important to note that, any moisture stress during the flowering phase may result into production of more staminate flowers in the seed parent, which is undesirable. Drip irrigation system may be coupled with adoption of black polythene mulch to conserve moisture, reduce weed growth, maintain soil temperature and reduction of damage to the fruits due to rotting. After laying of the mulch sheet, small punches are made at equal distances on the polythene sheets to sow the seed or to transplant seedlings.

Weed management: Though the pumpkin crop grows vigourously, clean cultivation is highly recommended. For effective management of weeds, frequent shallow cultivation is recommended. Although pumpkin is a deep and tap rooted plants, majority of the feeder roots are found to be confined near the surface. Thus, deep cultivation may cause serious damage to the feeder roots and is recommended to be avoided. Throughout the entire cropping season, three to four hand-weeding and hoeing is recommended. When manual labours are scare, application of pre-emergence herbicide like Basalin @ 2.0 lit/ha is recommended to manage the weeds before sowing or transplanting.

Floral biology and pollination: Pumpkin is a monoecious crop. Thus, high level of cross pollination is observed. The cross pollination is mainly carried out by honey bees. The anthesis of flowers start at 5.00 am and completes by 10 am. The anther dehiscence and peak receptivity of stigma also coincides with the same period. In natural condition, the staminate to pistillate flower ratio is 21:1. However, it can be narrowed down to 14:1-18:1 with the four foliar application of ethephon at 100-250 ppm, starting at two true leaf stage and at weekly intervals thereafter. In *C. pepo*, flower anthesis starts at 3.30 am and the dehiscence of anther takes place between 4.00 am to 11.00 am. The stigma remains receptive from 6.00 am to 2.00 pm while pollens remain viable up to 16 hrs after anthesis. The staminate flowers start emerging earlier than pistillate flowers, hence during the period of production of staminate and pistillate flowers, the presence of ample number of honey bees are very crucial for maximum fruit set, development and consequently high seed yield. Among different species of honey bees, *A. florae* is the most common. However, the species like *A. dorsata, A. cerena, A indica* and *A. mellifera* also contributes substantially to the fruit set. The bee activity is found to be maximum between 7.00 am to 9.00 am.

To maximize fruit set, placement of at least one beehive at the corner of the field is recommended for every hectare of area under seed production. It is also important to note that, for proper growth and development of the fruits and seed yield, every pistillate flower has to be foraged by flower has to be visited by at least 15 times by the honey bees. The bee hives should be placed three to five days after the appearance of the first blooms or when the 10-15% plants of the seed parent show anthesis of pistillate flowers.

Pollination method for hybrid seed production: In pumpkin, anthesis of flower occurs early in the morning at around 5.00 am and flowers remain open up to 10.00 am. Hence, pollination at early morning results with more success and seed production as after the peak anthesis hours, the pollen become wet and sticky. Since, pumpkin is a monoecious plant, it is easily emasculated by pinching of the staminate flowers before their anthesis in the seed parent. The seed production can be done both by natural pollination through honey bees in isolation as well as by hand pollination.

In natural pollination in isolation, utmost care should be taken to remove the staminate flowers from the seed parent before its anthesis to minimize the chances of self or sib pollination. It is a cheapest method of hybrid seed production as the pollination is carried out by the honey bees. Similarly in hand pollination, the staminate flowers from seed parent are removed regularly before the anthesis. The pistillate flowers that are likely to be opened next day in the seed parent, the corolla is tied with a thread during the evening hours. Similarly, the staminate flower bud of the pollen parent which is about to open in the next day is tied with a thread to prevent its anthesis. On the next day, the staminate flowers are collected from the seed parent and the whorl of corolla is removed to expose the anther column. The thread tied on the pistillate flower of the seed parent is also removed carefully to avoid any injury to the ovary. The anther column is then rubbed gently on the stigma of the pistillate flower flowed by tying of the corolla again or covering it with the perforated butter paper bags. As compared to the natural pollination in isolation, hand pollination in hybrid seed production of pumpkin ensures higher number of fruit set per plant, number of properly developed seeds per fruit that result into higher seed yield per plant.

Stage of rouging: Rouging is the process of removal of undesirable plants from the seed production plot to ensure both physical and genetic purity of the seeds. Thus, in hybrid seed production of pumpkin, rogueing at the following stages of crop growth becomes essential and is recommended.

1. Pre flowering stage: The plants of both seed and pollen parent are checked thoroughly for possible variation in leaf shape, size and colour. If any such plant is identified, it should be removed immediately.

2. Early flowering and fruit set stage: In this stage, rogueing is carried out to check the trueness to type of the developing fruits. If any fruits show undesired morphological characters, it should be removed immediately.
3. Fruit developing stage: The same process that is followed in early flowering and fruit set stage is also followed here.
4. Fruit maturity and harvesting stage: Only the well-developed and true to type fruits with the expected varietal characters are only selected for seed extraction. All the deformed and disease infected fruits should be discarded. If any plant shows the symptoms of pumpkin mosaic virus or mottle mosaic virus at any stage of plant growth, the plant should be uprooted and buried. No fruits from such plant should be selected for seed extraction.

Field standards: Pumpkin and Squash

Factors	Pumpkin		Squash	
	Maximum permitted (%)		Maximum permitted (%)	
	Foundation seed	Certified seed	Foundation seed	Certified seed
Off type in seed parent	0.01	0.05	0.05	0.05
Off type in pollen parent	-	0.05	-	0.05
Male flower shedding pollens on seed parent	-	0.05	-	0.10
Plants affected by seed bordering	-	0.10	0.10	0.50

Fruits regulation: The number of fruits per plant varies from variety to variety. However, the average fruit set ranges from 1.78 to 3.27 per plant within 45 days after the anthesis of first pistillate flower. The fruits set at the later stages remains small and yield less quantity of seeds per fruit. Thus, the fruits which are set through open pollination after the specific time should be removed regularly to maintain the favourable relationship between the source and sink. Further, this also reduces the chances of mechanical mixture. The number of fruits retained on the plants of the seed parents have a greater impact on the different seed quality attributes like thousand seed weight, germination percentage and seed vigour. However, it is also important to note that the regulation of fruits per vine has profound effect on seed yield i.e. more the number of fruits per vine, greater the seed yield and vice-versa.

Maturity and harvesting: The maturity and harvesting of fruits vary for vegetable purpose and seed production. About 35-40 days more time required to harvest fruits for seed production as compared to the ideal stage of harvesting for fresh consumption. The golden yellow-colored fruits are harvested for seed purpose unlike green stage for vegetable purpose, which usually takes about 50-55 days after the pollination of the pistillate flower. About 2-3 harvesting of golden yellow-colored fruits are done. The single harvesting results in over ripening and desiccation of fruits.

Seed extraction and drying: In general, pumpkin seed extraction is done by cutting the fruits with a knife and the seeds are scooped out from the pulp. Carrying out seed extraction over concrete floor is recommended to ensure hygiene of seed. The seed extraction may also be done on terpuline sheets. Before seed extraction, the freshly harvested fruits are kept in shade for 6-8 days for curing to enable quick and easy separation of seeds from the pulp. After harvest, the seeds should be washed in running tap water to separate the chaffy seeds, fruit pulp and mucilage. Then the seeds are shade dried for one day and then dried in a hot air oven by maintaining temperature of 38-41°C initially and then the temperature is reduced to 32°C-35°C. The oven drying is recommended to be continued till the seed moisture content comes down at or below 10% for storage in ambient conditions. The seed moisture content should be brought down to 6% is the seeds are intended to be stored in vapour proof container.

Yield: On average 35-50 tonnes fruits are harvested from one hectare area for fresh consumption purpose, which vary from variety to variety. However, the seed yield is greatly influenced by the genotype, management practices, pollination method followed, number of fruits retained per vine etc. In Pusa Hybrid-1, seed yield per fruit ranges from 40.08 g - 52.33 g and with an average of 83.66 g seed yield per plant. In natural pollination and hand pollination, the weight of thousand seeds recorded was 134.9 g and 148.08 g, respectively. On an average about 300-400 kg seeds /ha can be obtained.

Plant protection measures: Pumpkin is infested with many pests and infected with many diseases. The main pests causing serious damage to pumpkin are pumpkin beetles, cucumber beetles, aphids, squash vine borers, squash bugs and leaf-eating ladybird. To manage these pests, one-teaspoon Carbofuran (approximately 3 g) should be mixed per hill at the time of sowing. If infestation is observed at the later stages of crop growth, spray of lambda cyhalothrin @ 2 ml per litre of water is recommended. In case of occurrence of downy mildew, four alternative sprays of Dithane M-45 and Ridomil @ 2.5 gm per litre of water may be provided at fortnight intervals.

Pumpkin mosaic virus: The mosaic symptoms are characterized with chlorosis of terminal leaves with dark green older leaves. In advance stage, the leaf size reduced drastically and gives crowded, stunted shoots. Plants infected at an early stage yield no fruits after the infestation. The fruits, when infected, show irregular and raised blisters. Aphids are the vector of this disease which spread it from one plant to another. Till date, no varieties have been developed with resistance to this viral disease. Removal and burring of the infected plants are suggested to reduce the secondary infection among the plants. Spraying of systemic insecticide like imidachlopride @ 0.5 ml per litre of water effectively manages the population of aphids so that secondary spread gets restricted.

8

Watermelon

E. Sreenivasa Rao, Saheb Pal, D.C. Manjunatha Gowda and R.N. Thontadarya

[1]*ICAR-Indian Institute of Horticultural Research, Hesaraghatta Lake Bengaluru-560089, Karnataka*

[2]*ICAR-Indian Agricultural Research Institute, Gauria Karma, Barhi Hazaribagh-825405, Jharkhand*

[3]*Namdhari Seeds Private Limited, Uragahalli, Bidadi, Ramanagara, Bidadi-562109 Karnataka*

Watermelon (*Citrullus lanatus* (Thunb.) Matsum and Nakai) is an important cucurbit crop throughout the tropical and subtropical regions of the globe. It is an annual creeping vine and native to the Southern part of Africa, most probably the areas around the Kalahari Desert. Presently, more than 1200 cultivars have been developed and are being grown in nearly 122 countries with tropical and subtropical climates. Watermelon is mainly valued for its ripe 'pepo' fruits which are consumed as fresh-cut dessert or as juice. The fruits are a rich source of carotenoids, minerals, phenolic compounds, flavonoids, vitamin C and Citrulline, a non-essential amino acid. The red colour of the watermelon flesh is due to the presence of lycopene, the content of which is 40% more than that of tomato. Besides, there is good variation available for flesh colour ranging from red, pink, yellow, orange, white and green. Owing to the ability to prevent several chronic diseases due to the presence of pigments like lycopene, anthocyanin, carotene, xanthophyll etc., presently watermelon is also considered as a functional food. The exocarp (rind) is also consumed as stewed, pickled or stir-fried or may be used for the preparation of 'tooty-fruity'. Watermelon seeds are considered a good omen and are gifted during Chinese New Year in East Asia. The seeds are also used in the baking and confectionary industries as these contain a characteristic nutty flavour. The seeds are also rich in protein, vitamins as well as omega-3 and omega-6 fatty acids.

The varietal segments of watermelon are dynamic and evolve to the requirements of the market. Earlier, the large, round-fruited jubilee watermelon varieties/hybrids were preferred while presently, small and oblong-fruited varieties/hybrids are trending. However, among the various flesh colours available, the red-fleshed varieties are mostly preferred. Thus, both in terms of the quantity and value of the production, the crop has emerged as one of the most economically important food crops in the world (FAO, 2017). Globally, the watermelon seed market is predominated by hybrid varieties due to the advantages of high yield, better sweetness, intense flesh colour, uniformity of the fruit traits, transportability and tolerance/resistance to various biotic and abiotic stresses in such hybrids. The estimated value for global watermelon seeds market size was 507.70 million US Dollars in 2018 and the market size is also expected to increase by a CAGR of 5.8% during 2019-2025.

Ample genetic variability exists for different horticultural traits in watermelon but recent studies have reported narrow genetic diversity across cultivated varieties. This is probably due to the 'founder effect' arising from directional selection for a few consumer-appealing traits and the use of a few elite lines for the development of these varieties. Thus, there is an urgent need to broaden the genetic base of the parental inbred accessions by pre-breeding efforts. Development of such lines/inbreds with broad genetic base and their deployment in the breeding programme is the need of the hour for the development of hybrid varieties suitable for cultivation even under the changing climatic scenario.

Floral Biology of Watermelon

Predominantly, watermelon cultivars are monoecious with staminate and pistillate flowers on the same plant while some cultivars produce hermaphrodite flowers replacing the pistillate flowers to become andromonoecious. The ratio of the staminate to pistillate flowers can reach up to 17:1 and is dependent upon several factors *viz.*, the genetic makeup of the cultivar, age of the plant, prevailing temperature and photoperiodic conditions, application of nitrogenous fertilizers and availability of moisture etc. During the initiation of flowering, staminate flowers predominate followed by the production of pistillate flowers and at the end of the flowering period, the staminate flowers again predominate.

The staminate and pistillate flowers are solitary and are borne on the leaf axils. The inferior ovary of the flower remains attached to the pistillate flower buds. The long pedicels of the staminate flowers and the presence of ovaries in the pistillate flowers act as distinguishing features of staminate and pistillate flowers. The anthesis occurs between 5:00 AM and 9:00 AM. Depending upon the prevailing temperature and humidity, the flowers may remain fresh till the afternoon. The stigma becomes shiny and sticky and remains receptive from the time of anthesis till the dehiscence of the anthers. However, the prevalence of maximum receptivity is generally found between 6:00 AM to 10:30 AM. The staminate and pistillate flowers can bloom on the same day on the same plant, however, the staminate flowers tend to open before the opening of the pistillate flowers.

Breeding Objectives with the Current Priorities

The breeding objectives of watermelon are dynamic and change with time and region. However, the following are the objectives being pursued by breeders currently.

1. To develop strong vines with more number of branches.
2. To develop varieties that bear pistillate flowers at the lower nodes and uniform fruit ripening.
3. To develop varieties with the following desirable fruit quality traits to meet varied market and consumer preferences.
   a. Round or oblong fruit shape.
   b. Large, medium or small fruit size.
   c. Fruits with green, dark green or light green thick and tough rind with resistance to cracking with or without stripes.
   d. Firm red, canary yellow, pineapple yellow, orange or pink flesh with fewer and smaller seeds.
   e. Sweet flesh with more than 10°B TSS.
   f. Fruits with good transportability and shelf life.
4. Higher number of A-grade fruits per plant resulting in a higher profit to the farmer.
5. Resistance to diseases like Fusarium wilt, watermelon bud necrosis virus, gummy stem blight, anthracnose, downy mildew and powdery mildew and resistance to pests like fruit fly, fruit borers, aphids and red pumpkin beetle.

Varietal Achievements

Major Centres of Breeding and Varietal Achievements

In India, the first hybrid variety of vegetable crops was reported in bottle gourd during 1976, while hybrid watermelon breeding started in India during 1980s (Singh, 2000). The watermelon improvement in India is mainly confined to few major institutes like ICAR-Indian Institute of Horticultural Research, Bengaluru; ICAR-Indian Agricultural Research Institute, New Delhi; ICAR-Central Arid Zone Research Institute, Jodhpur; Punjab Agricultural University,

Ludhiana; Rajasthan Agricultural Research Institute, Durgapura and ICAR-Indian Institute of Vegetable Research, Varanasi. In the private sector, the major players in watermelon breeding are Syngenta Seeds, Seminis Seeds, Numhems Seeds, Namdhari Seeds, Kalash Seeds, Acsen HyVeg, US AgriSeeds, Sagar Biotech, Pahuja Seeds, East West Seeds, Chamunda Seeds, Sakata Seeds etc. Several public and private sector hybrids and varieties have been developed in watermelon by these organizations.

The major reasons for the popularity of F_1 hybrids in watermelon are,

- Very high genetic diversity: resulting in heterosis as high as 150% in many cases
- Majority of agronomic traits are controlled by over-dominance gene action: resulting in better heterosis
- No inbreeding depression: resulting in robust inbred parents and ease in maintenance of parental lines
- Higher number of seeds/pollination and low seed rate /unit area: leading to efficiency of seed production
- The potential yield advantage over open-pollinated varieties may reach as high as 33% -100% in watermelon and manifested as earliness, fruit size and fruit number per plant.

The salient features of varieties/hybrids developed and released from the public sector are presented below.

Name of the variety	Releasing organization/institute	Salient features
Asahi Yamato	ICAR-IARI, New Delhi	It is a mid-season variety with medium-sized ized-fruit ranging from 6-8 kg. It has a light green rind and deep pink flesh. TSS is about 11-13°B and it takes 95 days after sowing to reach maturity.
Sugar Baby	ICAR-IARI, New Delhi	It is a slightly small-fruited variety with round fruit with a bluish-black rind. It has deep pink flesh with small seeds. The average TSS is 11-13°B and it takes 85 days after sowing to reach harvest.
Pusa Bedana	ICAR-IARI, New Delhi	It is a triploid and seedless hybrid having aborted embryos and seeds. It was developed from the cross between Tetra-2 and Pusa Rasal. The fruits are somewhat irregularly triangular with thick dark green rind and deep pink flesh. The variety reaches harvestable maturity after 105 days of planting. However, the variety could not be popularized due to the costly and tedious seed production and difficulty in seed germination.

Arka Jyoti	ICAR-IIHR, Bengaluru	It is a variety developed through pedigree selection from a cross between IIHR-20 and Crimson Sweet. It is a round-fruited and mid-season variety with a light green rind with dark green stripes. The average fruit weight is 6-8 kg and the flesh has a crimson colour and 11-13°B TSS.
Arka Manik	ICAR-IIHR, Bengaluru	This variety has been developed through pedigree selection from a cross between IIHR-21 and Crimson Sweet. It has round to oval fruits with dark green and dull green stripes with deep red and very sweet flesh. The average TSS is 12-13oB and the average fruit weight is about 6 kg. The variety is suitable for long-distance transportation and possesses resistance to three diseases *viz.*, powdery mildew, downy mildew and anthracnose.
Arka Akash	ICAR-IIHR, Bengaluru	It is a high-yielding F_1 hybrid of IIHR-60-1 and Arka Manik. It has oval-round fruits with high TSS (11-12°B) and a yield potential of 90-100 tonnes per hectare.
Arka Aishwarya	ICAR-IIHR, Bengaluru	It is a high-yielding F_1 hybrid with round to oval fruits with dark green but discontinuous stripes. It has red crispy flesh and 13-14°B TSS. The average fruit weight is 7.5 kg and the yield potential is 75-80 tonnes per hectare. It has very good transport and keeping qualities.
Arka Madhura	ICAR-IIHR, Bengaluru	It is a triploid seedless watermelon hybrid derived from the cross between Tetra-1 × Arka Manik. It has round fruits with dark-green striped rind and crimson red flesh with a pleasant aroma. The TSS is 14oB. The average fruit size is kg and the yield potential is 50-60 tonnes per hectare. It takes 100-110 days after planting to reach harvest.
Arka Muthu	ICAR-IIHR, Bengaluru	It is an unique dwarf-vined, early maturing (75-80 days after planting) but high-yielding (55-60 tonnes/ha) variety suitable for high-density planting. The fruits are round to oval with dark green rind and deep red flesh. The average fruit weight is 2.5-3 Kg and the average TSS ranges from 12-14°B.
Arka Shyama	ICAR-IIHR, Bengaluru	It is an icebox segment open-pollinated variety with oblong fruits and dark greening black rind. The flesh is deep red, crispy and sweet (12°B TSS). It is an early variety and reaches harvest at 65-70 days of planting.
Improved Shipper	PAU, Ludhiana	This variety has large fruits with dark green rind. The average fruit weight ranges from 8-9 kg. However, the fruits are moderately sweet with a TSS of 8-9°B.

Special No.1	PAU, Ludhiana	It is an early maturing variety with round and small fruits with red flesh and red seeds. The variety is a little less sweet than the Improved Shipper variety.
Durgapura Meetha	ARS, Durgapura, Rajasthan	This is a late-maturing variety with round fruits. The rind colour is light green and the flesh is dark red and sweet. The average TSS is 11°B and the fruit weight is about 6-8 kg, the seeds have black tips and margins. The fruits mature in 125 days after sowing and have good keeping quality.
Durgapura Kesar	ARS, Durgapura, Rajasthan	It is a late-maturing variety. The rind colour is green with stipes and the flesh colour is yellow with moderate sweetness. The seeds are bold. The average fruit weight ranges from 4-5 kg.

Market Segments and Major Growing Belts

Based on the size of the fruits, the watermelon cultivars have been broadly grouped into ice box (2.5-3 kg), medium and big (7-10 kg) fruited varieties, while based on the shape they are classified as round and oblong types. Further, the cultivars with yellow and red flesh are available in the market. Based on rind pattern, the varieties may be classified as Sugar Baby (uniform green), Kiran (uniform black) or Jubilee (striped pattern) segments. The total market value of watermelon seeds in India is estimated at Rs. 191 crores. Personal-size watermelon or ice box segment has a market share of 41% by volume and 65% by value. This segment comprises the varieties with oblong fruit and red flesh and strong vigour of the plants bearing dark green or black fruits weighing about 3.0 kg each. The fruits are sweet with 11-12% TSS, crispy and firm flesh. This segment predominates in Andhra Pradesh, Karnataka, Maharashtra, Gujarat, Madhya Pradesh, Chhattisgarh, Uttar Pradesh, Punjab and Haryana. The leading varieties in this segment are Kiran, Sugar Queen, Arun, Suman, and H20. A new segment with similar characteristics of fruit except slightly larger fruit weight (4-6 kg) is also becoming popular including varieties like Maxx, Melody, Bahubali, Sagar King etc. The other family size segment includes relatively larger-sized fruit with striped rind ranging from 7-10 kg, both round and oblong with crispy flesh and relatively late maturity at 78-80 days after transplanting called the Jubilee segment. The Leading varieties/hybrids in this segment are Astha, Madhubala, NS295, Apoorva, Augusta, Ayesha, Madhav, Meghana, Suprit and Girish. These varieties are predominantly being grown in Karnataka, Tamil Nadu, Andhra Pradesh, Uttar Pradesh, Gujarat, Rajasthan, Odisha, Maharashtra and Chhattisgarh.

In India, watermelon is popularly grown in two major watermelon growing belts *viz.*, the Gangetic plains of northern India and the peninsular dry regions. In the Gangetic plains, Uttar Pradesh leads in production where Aligarh, Firozabad and Etah are the major producing districts. The major watermelon-growing districts of West Bengal are West Medinipur, Coochbehar and Murshidabad. In South India, the major watermelon-producing districts are Kolar, Mysore and Haveri in Karnataka; Chittoor, Anantapur and Kadapa districts of Andhra Pradesh; Villupuram, Kanchipuram and Thiruvallur districts of Tamil Nadu.

Ruling Hybrids in the Markets and the Reasons for their Popularity

The emergence of a strong private sector with an organized seed production and distribution system provided an impetus to hybrid development in watermelon. Different hybrids are popular across different watermelon growing belts based on the consumers' preferences as well as the end use of the fruits.

Major Watermelon Growing States	Leading Watermelon Hybrids	Brand / Seed Company	Segment
Andhra Pradesh	Apoorva	Seminis	Family Size-Crimson-Oblong- Red Flesh
	Madhubala	Nunhems	Family Size-Stripe-Oblong- Red Flesh (Jubilee)
	NS 777	Namdhari Seeds	Family Size-Stripe-Oblong- Red Flesh (Jubilee)
	NS295	Namdhari Seeds	Family Size-Stripe-Oblong- Red Flesh (Jubilee)
	MaxX	Nunhems	Personal Size-Sugar Baby-Oblong- Red Flesh
	Sagar King	Sagar Biotech	Personal Size-Sugar Baby-Oblong- Red Flesh
Bihar	Aastha	Nunhems	Family Size-Stripe-Oblong- Red Flesh (Jubilee)
	Madhuri	Nunhems	Family Size-Stripe-Oblong- Red Flesh (Jubilee)
Chhattisgarh	Augusta	Syngenta Seeds	Family Size-Sugar Baby-Round/Oblong- Red Flesh
	NS 750	Namdhari Seeds	Family Size-Stripe-Oblong- Red Flesh (Jubilee)
Gujarat	MaxX	Nunhems	Personal Size-Sugar Baby-Oblong- Red Flesh
	Sagar King	Sagar Biotech	Personal Size-Sugar Baby-Oblong- Red Flesh
Haryana	Kalia	Amazon Seeds	Family Size-Crimson-Oblong- Red Flesh
	Master 37	Chamunda Seeds	Personal Size-Sugar Baby-Oblong- Red Flesh
	Sagar King	Sagar Biotech	Personal Size-Sugar Baby-Oblong- Red Flesh

Karnataka	MaxX	Nunhems	Personal Size-Sugar Baby-Oblong- Red Flesh
	Melody	Kalash Seeds	Personal Size-Sugar Baby-Oblong- Red Flesh
	NS295	Namdhari Seeds	Family Size-Stripe-Oblong- Red Flesh (Jubilee)
	Sugar Queen	Syngenta Seeds	Personal Size-Sugar Baby-Oblong- Red Flesh
Madhya Pradesh	Augusta	Syngenta Seeds	Family Size-Sugar Baby-Round/Oblong- Red Flesh
	Sagar King	Sagar Biotech	Personal Size-Sugar Baby-Oblong- Red Flesh
Maharashtra	MaxX	Nunhems	Personal Size-Sugar Baby-Oblong- Red Flesh
	Melody	Kalash Seeds	Personal Size-Sugar Baby-Oblong- Red Flesh
	Sagar King	Sagar Biotech	Personal Size-Sugar Baby-Oblong- Red Flesh
	Sugar Queen	Syngenta Seeds	Personal Size-Sugar Baby-Oblong- Red Flesh
Odisha	Augusta	Syngenta Seeds	Family Size-Sugar Baby-Round/Oblong- Red Flesh
Punjab	Master 37	Chamunda Seeds	Personal Size-Sugar Baby-Oblong- Red Flesh
	MaxX	Nunhems	Personal Size-Sugar Baby-Oblong- Red Flesh
	NS295	Namdhari Seeds	Family Size-Stripe-Oblong- Red Flesh (Jubilee)
	NS34/Mishri	Namdhari Seeds	Personal Size-Sugar Baby-Oblong- Red Flesh
Rajasthan	Aastha	Nunhems	Family Size-Stripe-Oblong- Red Flesh (Jubilee)
	Cassatta	Acsen Hyveg	Family Size-Crimson-Oblong- Red Flesh
	Kalia	Amazon Seeds	Family Size-Crimson-Oblong- Red Flesh
	Lalpasand	Metahelix Veg Seeds	Family Size-Crimson-Oblong- Red Flesh
	NS295	Namdhari Seeds	Family Size-Stripe-Oblong- Red Flesh (Jubilee)

Tamil Nadu	Apoorva	Seminis	Family Size-Crimson-Oblong- Red Flesh
	Dragon King	Syngenta Seeds	Family Size-Stripe-Oblong- Red Flesh (Jubilee)
	MaxX	Nunhems	Personal Size-Sugar Baby-Oblong- Red Flesh
	NS295	Namdhari Seeds	Family Size-Stripe-Oblong- Red Flesh (Jubilee)
	Suman	Pahuja Seed	Family Size-Sugar Baby-Round/Oblong-Red Flesh
	Sweet Dragon/ Pakeeza	Nunhems	Family Size-Stripe-Oblong- Red Flesh (Jubilee)
Telangana	Andaman	East West Seeds	Family Size-Sugar Baby-Round/Oblong-Red Flesh
	Apoorva	Seminis	Family Size-Crimson-Oblong- Red Flesh
	NS295	Namdhari Seeds	Family Size-Stripe-Oblong- Red Flesh (Jubilee)
	Sagar King	Sagar Biotech	Personal Size-Sugar Baby-Oblong- Red Flesh
	SW 2208	SeedWorks	Personal Size-Sugar Baby-Oblong- Red Flesh
Uttar Pradesh	Aastha	Nunhems	Family Size-Stripe-Oblong- Red Flesh (Jubilee)
	Madhuri	Nunhems	Family Size-Stripe-Oblong- Red Flesh (Jubilee)
	Master 37	Chamunda Seeds	Personal Size-Sugar Baby-Oblong- Red Flesh
	MaxX	Nunhems	Personal Size-Sugar Baby-Oblong- Red Flesh
	SV5061WL	Seminis	Family Size-Stripe-Oblong- Red Flesh (Jubilee)
West Bengal	NS23	Namdhari Seeds	Personal Size-Sugar Baby-Oblong- Red Flesh
	Sagar King	Sagar Biotech	Personal Size-Sugar Baby-Oblong- Red Flesh
	Sato	Sakata Seeds	Personal Size-Sugar Baby-Oblong- Red Flesh

Seed Production Methodology

Pollination Behaviour and Crossing Methodology

Monoecy is the most predominant sex form in watermelon like most other cucurbits. Monoecy refers to the condition when both the staminate and pistillate flowers are borne on different nodes of the same plant. Pollination refers to the movement of the pollen from the anthers of the staminate flowers to the stigma of the pistillate flowers. Owing to the showy yellow petals, monoecious flowering habit, production of large and sticky pollen, presence of nectar glands at the base of the corolla, the pollination of watermelon is mainly mediated by honey bees. Further, the male flowers produce more nectar as compared to the female flowers to attract more honey bees. Therefore, seed production of open-pollinated varieties can be done by keeping 5-6 honey bee boxes/acre and maintaining an isolation distance of 1500m for foundation seed and 1000 m for certified seed. However, since there is no effective pollination control mechanism available in watermelon, the commercial practice of hybrid seed production is to do hand pollination. Thus, for every 3-4 rows of seed parent, one row of male parent is planted. The staminate and pistillate flower buds of male and female parents respectively that are about to open in the next morning are covered with cotton or butter paper bags. On the next day morning, bagging/cotton of the pistillate flowers is opened and from the staminate flowers, the pollen is transferred to the stigma of the pistillate flowers followed by bagging and tagging. Utmost care must be taken not to touch the stigmatic surface during bagging and/or pollination as any injury leads to the production of ethylene resulting in drop of the flower. The best time for carrying out hand pollination is during the morning between 7:00 am to 9:00 am.

Fig: Hand Pollination method

Method of Seed Production Being Practiced by Private Companies

Seed production at a commercial scale is being practiced in some definite pockets of Karnataka through contract farming with the farmers. Koppal district, Sira region of Tumkur district and Ranebennur region of Haveri district of Karnataka are considered as the hubs for commercial seed production of watermelon. The soil type in those regions is sandy loam, leading to early harvest of quality seeds. For contract farming, comparatively poor but interested farmers are selected and are allotted a plot (approximately 1.25 acres) or a maximum of up to two plots to maintain the quality of the seed lot. In every plot, approximately 5000 plants, including 4000 female and 1000 male plants are planted. Due to the non-availability of a cost-effective pollination control mechanism, hand pollination is practiced. For this purpose, about 8-10 labourers are engaged per acre for 10 days at the peak pistillate flowering phase of the plants of the seed parent. The cost of production thus reaches to about Rs. 50,000 per acre. After seed harvest, the seed lot is subjected to the Grow Out Test and if the lot passes the test, the farmers are paid 30% of the production cost. The small-seeded icebox varieties generally yield about 7-8 gm seeds per fruit or 30 kg seeds per acre while a jubilee variety with medium-sized seeds yields about 15 g seeds per fruit and 60 kg seeds per acre. The bold-seeded Jubilee variety yield about 20 gm seeds per fruit and 100 kg seeds per acre. The harvested seeds are collected by treating them with a commercial product anti-microbial product named Tsunami 100 @ 12.5 ml per litre of water for 15 minutes to prevent the growth of pathogenic microorganisms during the controlled fermentation process of the pulp.

Seed Production of Triploid Seedless Watermelon

- The seedless triploid watermelons are also F_1 hybrids of the cross between tetraploid and diploid inbreds.
- For producing tetraploids, the young seedlings of the diploid genotypes are treated with an aqueous solution of colchicine (0.2 - 0.4 percent). The tetraploids can be maintained by selfing.
- The tetraploid is pollinated with the pollen from the diploid parent to produce triploid seed.
- The triploid is highly sterile and seedless but produces white rudimentary seeds or in some cases with colored empty seed coat.
- Generally, the small-seeded cultivars in which the rudimentary seeds are not so unattractive as in large-seeded varieties are preferred as parents to produce triploids.

IARI hybrid, Pusa Bedana and IIHR hybrid, Arka Madhura are examples of seedless watermelons developed in the public sector. Recently, KAU has developed Shonima (red fleshed) and Swarna (yellow fleshed) triploid seedless hybrids. In the private sector, Syngenta is trying to popularise its seedless hybrid in the name of 'Happy Family' in India and Nunhems hybrid is named 'Style'.

However, seed production of these triploids has been difficult due to poor germination and less vigour of tetraploid parents. Therefore, IIHR has developed a tissue culture protocol for multiplication of its seedless hybrid, Arka Madhura.

Fig: Arka Madhura

Cultural Practices for Hybrid Seed Production of Watermelon

- **Selection of Field and Preparation of Land**

  Fields with well-drained and fertile soil should be selected for seed production. The areas with a history of soil-borne diseases like Fusarium wilt and gummy stem blight should be avoided. The land should be ploughed deep followed by harrowing or by using a rotavator to fine tilth.

- **Planting**

  For seed production, the parents are planted at a lower density than commercial fruit production. This is specifically done to reduce inter-plant competition and encourage branching of the plants leading to higher seed yield and better quality. Further, wider spacing ensures a reduction in the severity of foliar diseases by slowing down the spread of secondary infection. The ideal row-to-row spacing is 1.5 m and plant-to-plant spacing is 0.75 m. The planting of the seed and pollen parent should be done at a ratio of 4:1 to have enough pollen availability for pollination. Staggered planting of the pollen parent may be done to ensure the availability of sufficient pollen during the peak pistillate flowering phase in the seed parent. An isolation of 5 m between blocks of the parental lines in case seed parent and pollinator are planted in separate blocks and hand pollination is to be adopted.

- **Pollination**

  Watermelon is a cross-pollinated crop and pollination is mainly mediated through honeybees. For seed production of open-pollinated varieties, honey bee colonies are kept @ 15 per hectare for effective pollination. However, manual pollination is done in case of hybrid seed production.

- **Irrigation and Aftercare**

  The watermelon plants for seed production are generally planted on raised beds with plastic mulch and drip irrigation. Need-based irrigation must be provided to maximize seed yield. After pollination of the seed parent, the plants of the pollen parent along with the set fruits (if any) may be removed to prevent chances of mechanical mixture. Thus, this practice helps in the maintenance of a high level of genetic purity in the seed lot.

Special Cultural Operations for Hybrid Seed Production of Watermelon

For better seed set and development, the plants from the seed parent are initially trained to have only two vines per plant and one fruit is allowed to be set on each vine.

Rouging

Rouging refers to the removal of off-type and diseased plants from the seed production field. Generally, four or at least three rouging are recommended in watermelon seed production. First Rouging is done before the commencement of flowering to rouge out the undesirable plants based on the vegetative characteristics. Second rouging is recommended at the early flowering stage to confirm the true to type of developing fruits. The third rouging is recommended at the fruit-developing stage and the fourth rouging is recommended to be done at the fruit harvest stage.

Seed Harvest

In watermelon, the harvest or horticultural maturity coincides with seed maturity. However, the fruits meant for seed purposes should be left in the field for at least seven days after reaching harvest maturity. Generally, depending on the varieties, it takes 40-60 days after pollination to reach harvest maturity depending on the environment. The harvest maturity can be determined by observing the drying of the tendrils opposite to the node bearing the fruit, the development of a pale-yellow colour at the underside of the fruit surface touching the ground and a dull sound on tapping the fruits. The harvested fruits are sorted and the fruits with the best size, appearance and quality are selected for seed extraction.

Seed Extraction of Methods

Unlike most other cucurbit vegetable crops, watermelon shows parietal placentation and the seeds remain distributed throughout the placenta or the central cavity of the fruit. Therefore, the hybrid seeds can be hand-extracted by fermentation of the pulp for 24 hours followed by washing of the seeds and drying. The pulp along with the seeds is firstly taken into drums/polypropylene bags and packed airtight to create an anaerobic condition for the next 24 hours. After that, the pulp is macerated on a stainless-steel screen under running tap water. Thus, the pulp gets easily separated from the seeds while washing. The floating seeds indicate immaturity and are discarded. The seeds which sink at the bottom, are collected and dried in the sun for two days followed by shade drying to bring down the moisture content up to 7% for storage in ambient conditions and to 6% for storage in vapour-proof conditions. However, utmost care must be taken during the fermentation process as over-fermentation may lead to discolouration of the seed coat and may bring down the germination percentage.

Seed Processing and Machinery

In large-scale seed production, seed extraction can be done by mechanical bulk seed extractors. The fruits are fed into the machine which separates the rind, pulp and seed. The seeds can be washed in washing screens and dried in rotary driers or under shade. Grading of the seeds can be done based on their density using a specific gravity separator.

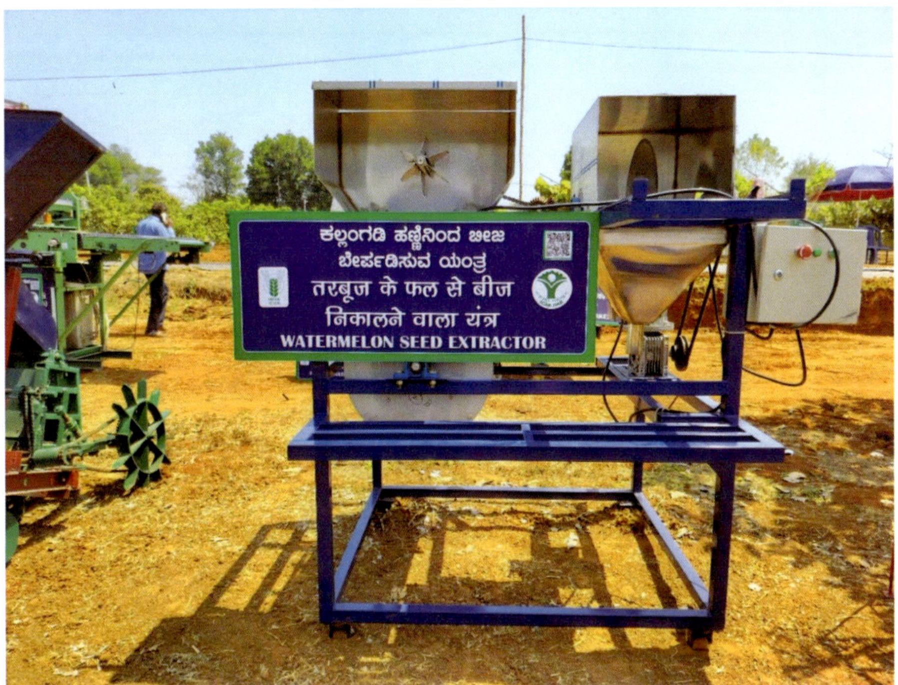

Fig: Power-operated watermelon seed extractor developed by ICAR-IIHR, Bengaluru.

Seed Yield

Seed yield is a factor of varietal character of seed parent, seed size (test weight), management practices followed and pollination efficiency. Generally, the varieties with longer duration (late varieties) have bold seeds and produce more yield compared to early varieties. Under good management practices, a yield of 400-500 kg seeds per hectare is obtained.

Field and Seed Standards of Watermelon: The permissible limits as per law for watermelon seed are as follows:

Field Standards

Factor	Maximum Permissible Limits (%)	
	Foundation Seed	Certified Seed
Presence of off-types in the seed parent	0.010	0.050
Presence of off-types in the pollen parent	None	0.050
Presence of shredders in the seed parent	None	0.10
Objectionable and crossable weed plants (eg. wild watermelon plants)	None	None

Seed Standards

Factor	Class of Seed	
	Foundation Seed	**Certified Seed**
Pure seed (minimum)	98.00%	98.00%
Presence of inert matter (maximum)	2.00%	2.00%
Presence of other crop seeds (maximum)	None	None
Presence of weed seeds (maximum)	None	None
Objectional weed seeds (maximum)	None	None
Presence of seeds of other distinguishable varieties (maximum)	Five/Kg	Ten/per Kg
Germination percentage (minimum)	60.00%	60.00%
Moisture content for ambient storage (maximum)	7.00%	7.00%
Moisture content for vapour-proof container (maximum)	6.00%	6.00%

Varietal Protection and DUS Testing Guidelines

In India, the Protection of Plant Varieties and Farmers' Rights Authority (PPVFRA) grants protection for varieties against commercial misuse. Registration with PPVFRA shall be done based on DUS (Distinctness, Uniformity and Stability) testing. For watermelon, this is being carried out at two centers namely, CIAH, Bikaner and IIHR, Bangalore. The testing is done as per guidelines available on the PPVFRA website and published in Plant Variety Journal, Vol-8 (7), 2014. The testing shall be done in two independent but similar growing seasons (generally summer for watermelon) in these centers. Each entry shall be tested along with reference varieties in three replications of 35 plants each. The data shall be recorded on 10 plants in each replication. A total of 27 characters shall be recorded in the following stages

1. **Seedling stage**: Shape of cotyledon
2. **Vegetative characters**: Length of internodes, Length of leaf blade, Width of the leaf blade, Leaf blade colour, Degree of lobing of the primary leaf blade, Degree of lobing of the secondary leaf blade, Petiole length
3. **Flowering characters:** Appearance of first pistillate/ perfect flower in 50% plants from date of sowing, Sex expression, Male Sterility, Length of the ovary, Width of the ovary, Ovary pubescens, Ovary colour
4. **Fruit characters**: Fruit shape in longitudinal section, Colour of the fruit rind, Fruit grooves, Fruit stripes, Fruit length, Fruit diameter, Fruit size Colour of flesh, Number of seeds per fruit, Seed length, Seed width Seed coat colour

The data collected for two seasons and at two locations will be compiled and analyzed by PPVFRA examiner to check for Distinctness, Uniformity and Stability of the entry. If the entry clears the test, it shall be granted a "Certificate of Registration" with a protection period of 15 years against commercial misuse.

9

Muskmelon

D.C. Manjunatha Gowda, E. Sreenivasa Rao and Anil Kumar Nair
Division of Vegetable Crops, ICAR-Indian Institute of Horticultural Research
Hesaraghatta Lake Post, Bengaluru-560089, Karnataka

Melons are annual plants that have up to three meters of climbing, creeping, or trailing vines. Several of them are consumed as dessert fruits while a few varieties/types are used for vegetable purposes. Generally, the fruits of commercial varieties are sweet, juicy, delicious with good nutritional and medicinal properties and a daily intake of 120-150 g can help in weight loss, better eye and skin health (Shivapriya *et al.*, 2021).

Currently, muskmelon is being cultivated throughout the world, especially under tropical and subtropical climatic conditions. In India, muskmelon is grown in 54,000 ha with a production of 1.23 million tons (NHB, 2018). The estimated national seed requirement per year is about 27 tons with a value of about Rs. 27 crores. The crop is cultivated mainly in *rabi* season (November to February sowings), thus limiting its availability in the market during summer. Rarely, melons are cultivated for the off-season market, and those fetch a premium price. The total value of muskmelon fruits in the domestic market is estimated as Rs.1231 crores.

Origin and Distribution

The place of origin of this large and polymorphous species is not known with absolute certainty. 40 or more wild species of *Cucumis* are found to grow in the tropics and subtropics of Africa and it is generally considered that Africa as the possible centre of origin for this crop. Although melon was introduced into Asia at a comparatively later date, there are undoubtedly well-developed secondary centers of origin of *C. melo* in India, China and the Southern USSR. Dhillon *et al.* (2012) reviewed the melon landraces of India. Three main types of non-sweet melons are cultivated as landraces in India *viz.*, *C. melo* ssp. *flexuosus*, ssp. *acidulous* and ssp. *momordica*. The fourth group is formed by a semi-domesticated melon known as 'chibber' or 'kachri' used for pickles and chutneys which is referred to as *C. melo* ssp. *agrestis* by few and ssp. *acidulous* by others. All of these are intercrossable and also crossable to *C.*

melo subsp. *melo*. In South India, the Western Ghats and river beds of Kaveri and Krishna are home to rare and variable oriental pickled melon (*Cucumis melo* var. *conomon*) and snap melon (*C. melo* var. *momordica*) landraces. The Indian gene pool has gained a lot of relevance in global breeding programs as a major source of resistance to several biotic stresses including *Fusarium* wilt, downy mildew, powdery mildew, aphids and viruses.

Classification of Melons

Cucumis melo is classified into two sub-species *viz.*, subsp. *melo* and subsp. *agrestis*, based on the length of the ovary's pubescence (Jeffrey 1980). Based on flower and fruit characteristics like sex expression, fruit size and shape, sweetness, flavour, color and climacteric attributes, these two subspecies are further classified into seven horticultural groups (Munger and Robinson 1991; Robinson and Decker-Walters 1997; Pitrat *et al.* 2000).

Later Pitrat (2008) and Pitrat *et al.* (2017) proposed European taxonomy, known as "infraspecific taxa," which is largely recognized and divides domestic melons including Asian varieties into 19 distinct horticultural groups.

Eleven groups for subsp. *melo*

1. *Cucumis melo* L. subsp. *melo* var. *adana*
2. *Cucumis melo* L. subsp. *melo* var. *ameri*
3. *Cucumis melo* L. subsp. *melo* var. *cantalupensis*
4. *Cucumis melo* L. subsp. *melo* var. *casaba*
5. *Cucumis melo* L. subsp. *melo* var. *chandalak*
6. *Cucumis melo* L. subsp. *melo* var. *chate*
7. *Cucumis melo* L. subsp. *melo* var. *chito*
8. *Cucumis melo* L. subsp. *melo* var. *dudaim*
9. *Cucumis melo* L. subsp. *melo* var. *flexuosus*
10. *Cucumis melo* L. subsp. *melo* var. *inodorus*
11. *Cucumis melo* L. subsp. *melo* var. *ibericus*
12. *Cucumis melo* L. subsp. *melo* var. *indicus*
13. *Cucumis melo* L. subsp. *melo* var. *reticulates*
14. *Cucumis melo* L. subsp. *melo* var. *tibish*

Five groups for subsp. *agrestis*

15. *Cucumis melo* L. subsp. *agrestis* var. *acidulous*
16. *Cucumis melo* L. subsp. *agrestis* var. *chinensis*
17. *Cucumis melo* L. subsp. *agrestis* var. *conomon*

18. *Cucumis melo* L. subsp. *agrestis* var. *makuwa*
19. *Cucumis melo* L. subsp. *agrestis* var. *momordica*

Chromosome Number and Genomics

The muskmelon crop is a diploid with 2n = 24. The first comprehensive list of known melon genes was published in 2006 (Pitrat, 2006), and then an updated list was published in Cucurbit Genetics Cooperative by Catherine (2011). The first melon genome sequence was published in 2012 (Garcia *et al.*, 2012) which paved the way for an in-depth understanding of genetic control of several traits in melons including fruit quality, plant architecture and insect/disease resistance.

Breeding Objectives

1. High quality fruit
    - High TSS (> 12°Brix)
    - High average fruit weight resulting in high yield
    - Strong rind with a pleasing exterior colour
    - Uniform netting all over the fruits in case of cantaloupes
    - Thick flesh with attractive colour
    - Crisp flesh texture
    - Long shelf life
2. Disease resistance: powdery mildew, downy mildew, viruses (cucumber green mottle mosaic virus, cucumber, mosaic virus, squash mosaic virus)
3. Adaptation to specific environments.

Breeding methods: Mass selection, pedigree selection and heterosis breeding methods are the most frequently used melon improvement methods followed in India. A list of varieties developed and notified in India are as follows:

Sl. No.	Name of Variety/Hybrid	Breeding method	Notification details
1	Kashi Madhu	Mass Selection	CVRC
2	Pusa Sarbati	Pedigree method	CVRC
3	Hara Madhu	Mass Selection	CVRC
4	MHY-5	Mass Selection	CVRC
5	Pusa Madhuras	Mass Selection	CVRC
6	Arka Rajhans	Mass Selection	CVRC
7	Arka Jeet	Mass Selection	CVRC
8	Durgapura Madhu	Mass Selection	CVRC
9	NDM-15	Mass Selection	CVRC

10	Pusa Rasraj	F_1 hybrid	CVRC
11	Punjab Sunehari	Pedigree method	SVRC
12	Punjab Rasila	Mass Selection	SVRC
13	Hisar Madhur	Mass Selection	SVRC
14	RM-43	Mass Selection	SVRC
15	MHY-3	Mass Selection	SVRC
16	RM-50	Mass Selection	SVRC
17	Punjab Hybrid-1	F_1 hybrid	SVRC
18	MH-10	F_1 hybrid	SVRC

CVRC: Central Variety Release Committee; SVRC: State Variety Release Committee

Heterosis Breeding

The major reasons for the popularity of F_1 hybrids in melons include,

- Very high genetic diversity among the parental inbred lines results in a very high magnitude of heterosis reaching up to 150% in many cases.
- Majority of the agronomic traits show over-dominance gene action: resulting in higher heterosis.
- Negligible inbreeding depression: resulting in robust inbred parents and ease in maintenance of parental lines.
- Higher number of seeds/pollination and low seed rate /unit area leads to efficient use of hybrid varieties.
- Availability of pollination control mechanisms like male sterility, gynoecy, monoecy and also amenable to chemical sex modification enables the production of hybrid seed in a cost-effective manner.
- Predominance of entomophily enables the efficient transfer of pollens and saves cost of manual pollination.

Munger (1942) was the first to exploit hybrid vigour in muskmelon. Heterosis as high as 313.6% in muskmelon (Mishra and Seshadri, 1988) has been recorded. However, in general, the potential yield advantage over open-pollinated varieties may reach as high as 33% -100% in melons.

Manifestation of heterosis: The successful use of hybrids mainly depends upon the existence of an economically significant level of heterosis, ease of cross-pollination and seed set per pollination to make hybrid seed production economical and also an efficient system for the maintenance of parental lines. Heterosis in muskmelon is generally expressed for total yield (20-100%) and earliness (3%), average weight of first three fruits, soluble solids/sugar content, fruit shape and degree of netting.

Technique for Hand Pollination

- In muskmelon most of the cultivated varieties are andromonoecious bearing hermaphrodite and staminate flowers on the same plant while some genotypes are monoecious bearing both staminate and pistillate flowers on the same plant but at separate nodes.
- In the seed parent, the perfect flowers in andromonoecious lines are emasculated and covered with butter paper bags about 24 hours before anthesis. For emasculation, anthers are removed as a group or anther column, with or without the surrounding corolla, by griping the base of the anthers and petals with forceps, which are then removed by a firm but steady pull.
- In the pollen parent, the corolla of the staminate flower bud is tied with string or very thin wire or cotton plug to prevent it from opening instead of covering them with a butter paper bag.
- The following day when the flowers open, the staminate flowers are collected from pollen parent and pollens are dusted directly onto the stigma of the emasculated flowers of seed parent.
- After pollinating, the pistillate flower of the seed parent is again covered with a butter paper bag and tagged with a small label on which the details of the cross and date of pollination are written with pencil.
- It is better to remove the paper bag after 4-5 days of pollination to minimize the chances of build-up of humidity around the growing ovary to obtain a better fruit set and size.

Actions to Enhance the Seed Set Include

1. Only one or two primary branches should be kept per plant to maximize production of pistillate flowers and to ensure a better fruit set.
2. It is also advisable to pollinate the first or second hermaphrodite flower buds opened on a vine called crown set, as in later stages the fruit setting percentage gradually reduces.
3. Only two to four perfect flowers on each plant may be pollinated and the vine growth tips may be removed after successful pollination and fertilization to minimize the production of further flowers that cause diversion of plant energy.
4. The other hermaphrodite flower buds or other fruits set on the same vine on which the flowers were pollinated should also be removed to prevent energy diversion.

(a) (b) (c)

Fig: Hand pollination in muskmelon, **a)** Emasculation of seed parent, **b)** preparation of staminate flower for pollination, **c)** Hand pollination

Methods of commercial hybrid seed production: In muskmelon the following procedures may be adopted for producing F_1 hybrid seeds.

1. **By artificial pollination**: Bagging of emasculated perfect flowers followed by hand pollination on next day can be followed as explained in the previous section. This is the most commonly practiced method currently in the private sector.
2. **Removal of staminate flowers and use of insect pollination**.

   In situations where the seed parent is monoecious in nature, there shall be no need of emasculation. Therefore, Curtis (1939) suggested planting the seed parent and pollen parent rows alternately and the removal of staminate flowers from the plants of the seed parent until one or two fruits have been set on the plants of the seed parent.
3. **Use of morphological markers like leaf lobing**
   - The non-lobing leaf character is governed by a single recessive gene which is expressed in the seedling stage and can be used as a marker gene for producing hybrid seeds, as suggested by Mohr *et al.* (1955).
   - The inbred lines of non-lobed and lobed may be planted in alternate rows at a spacing of 1 m × 2 m and the hybrid seeds should be collected from the plants with non-lobed leaves only.
   - The seedlings having lobed leaves are the true F_1s and others are the resultant of self-pollination. They can be easily distinguished in the nursery and rouged out.
   - In directly sown crop, Mohr. *et al.* (1955) suggested sowing about 6 to 8 seeds per hill as on average as only about one-third of the plants will be pure F_1 hybrids.

- According to Whitaker and Davis (1962) this method may be quite easy and economical for the production of hybrid seeds.
- Yellow leaf and glabrous seedlings are other monogenic recessive characters that can be used as seedling markers in muskmelon.

4. **Use of pollination control mechanisms**: Genic male sterility (*ms1* to *ms5* genes) has been reported in muskmelon (Bohn and Whitaker, 1949; Bohn and Principle, 1964), which can be exploited for the production of F_1 hybrid seeds. However, male sterility has not been used for commercial production of hybrid seeds yet because the stability of male sterile lines under different environments is a major concern. Other pollination control mechanisms in muskmelon include Monoecy (M1-4) and gynoecy (WI998). A few examples of hybrids developed in India employing pollination control mechanisms are as follows:

   a. Pusa Rasraj = M3 (Monoecious) × Durgapura Madhu

   b. Punjab Hybrid-1 = (Male sterile) × HaraMadhu

   c. MH-10 = WI998 (gynoecious) × Punjab Sunehri

5. **Chemical sex modification**

Chemical	Stage of application	Effect
Ethrel @ 150-200 ppm	2-4 true leaf stage and before flowering	Pistillate flower comes first and results in higher fruit set
Silver thio sulphate @ 300-400 ppm and GA_3 1500-2000ppm	2-4 leaf stage or any active growth stage	Induces production of hermaphrodite flowers in the gynoecious lines

Harvesting and Seed Extraction

- Fruits of climacteric type hybrids are harvested after full maturity (full slip stage) when the fruit turns into yellow colour and gets separated from the vine automatically.
- In non-climatic varieties that do not separate from the vine upon ripening, harvesting should be done when cracks develop at the junction point of fruit peduncle. Such fruit requires curing for 7-10 days in a well-ventilated shade before seed extraction.
- As the seeds are present in the central cavity, they are extracted by hand and kept for fermentation for 24 hr for separation of seeds from the placental tissues and washed with clean water.
- The seed is dried under shade for 3-4 days to bring down the moisture content below 8% for storage under ambient conditions.

- The F_1 hybrid seed yield ranges from 300-400 kg/ha depending upon the variety, nutrition etc. and management practices followed.

Field and Seed Standards: The permissible limits as per law for muskmelon hybrid seed certification are as follows:

A. **Field Inspection**: Four inspections must be performed at least: before, during flowering and fruiting, and full fruit stage and before harvesting.

B. **Isolation distance**: 1500 m for foundation seed and 1000 m for certified seed.

C. **Field Standards:**

Factor	Maximum Permissible Limits (%)	
	Foundation Seed	Certified Seed
Off-types, objectionable weeds, or diseased plants in seed parent.	0.010	0.050
Off-types, objectionable weeds, or diseased plants in pollinator parent.	None	0.050
Weedy melons and objectionable plants.	None	None
Plants infected with diseases spread by seeds the disease transmitted by seeds is the Cucumber Mosaic Virus (CMV).	0.10	0.20
Pollen shedding cymes, or fertile segregants, in the seed parent.	0.050	0.10
If an andromonoecious line is utilized, the male flowers shed pollen in the seed parent.	None	0.10

D. **Seed Standards**

Factor	Class of Seed	
	Foundation Seed	Certified Seed
Pure seed (minimum)	98.00%	98.00%
Presence of inert matter (maximum)	2.00%	2.00%
Presence of other crop seeds (maximum)	None	None
Presence of weed seeds (maximum)	None	None
Objectional weed seeds (maximum)	None	None
Presence of seeds of other distinguishable varieties (maximum)	Five/Kg	Ten/Kg
Germination percentage (minimum)	60.00%	60.00%
Moisture content for ambient storage (maximum)	7.00%	7.00%
Moisture content for vapour-proof container (maximum)	6.00%	6.00%

DUS Guidelines and Registration with PPV& FRA

DUS test guidelines for muskmelon have been notified by the PPVFRA in 2014. Two centers namely, ICAR-CIAH, Bikaner and ICAR-IIHR, Bengaluru have been identified for DUS testing of this crop. 100 g seeds of each entry should be submitted for testing. The testing shall be done in two independent but similar growing seasons (generally summer for muskmelon) in these centers. Each entry shall be tested along with reference varieties selected based on six grouping characters namely sex expression (at full flowering) Fruit: shape in longitudinal section, Fruit: color of the rind, Fruit: sutures Fruit: surface netting and Fruit: color of flesh. The test shall be conducted in three replications of 35 plants each. The data shall be recorded on 10 plants in each replication. A total of 34 characters shall be recorded in various stages of the crop including seedling, vegetative, flowering, fruit and seed characters. The data collected for two seasons and at two locations will be compiled and analyzed by the PPVFRA examiner to decide on the grant of Registration. An open-pollinated variety of muskmelon *viz.*, Arka Siri from IIHR was registered with PPV&FRA.

Agronomy for Muskmelon Seed Production

Soil type: Sandy and sandy loam soils with good drainage are the best. Poorly drained soils should be avoided for seed production of muskmelon.

Season of transplanting: It is generally sown during the months of November and December (for the *rabi* season) and February (for the summer season), recent trend is to sow to catch the Ramzan festival market.

Spacing: A row-to-row spacing of 1.5-1.8 m and a plant-to-plant spacing of 30-40 cm is recommended for muskmelon seed production.

Seedling Requirement: 6500-7000 plants/acre

Seed weight: 100 seed weight: 2.6 g. Generally, 200-250 gram seeds are required per acre.

Nursery raising: Pro-tray method: raised in a covered structure, trays are loaded with cocopeat that has been enriched by beneficial microbes. Seedlings can be drenched with bavistin (2g/L) or copper-oxy-chloride or Copper hydroxide (2g/L) on 10th day after sowing as a precautionary measure against damping off disease. When we are sure of a good germination percentage of the seed lot, direct sowing can be practiced. The crop shall be harvested 1 week early in such cases.

Seedling age: Seedlings will be ready (with one true leaf) for transplanting by 14-16 days after sowing, depending upon the prevailing temperature conditions.

Land preparation main field: Bring soil to a fine tilth after 2-3 ploughing and harrowing. A raised bed can be 15 cm high, 60 cm wide, and of a convenient length, depending on the drip system.

Enrichment of FYM: FYM enrichment has to be done in a partially shaded place. It takes around 21 days before field application. Add 1 kilogram of *Trichoderma* and *Psuedomonas* each with 1 ton of FYM. Keep crop wastes or gunny bags 20-30% moist to cover the heap. Sprinkle water when required. Mix one ton of enriched FYM with nine tons of FYM for the final field application.

Neem cake application: Apply neem cake @ 250 kg before final bed preparation. A portion of neem cake can also be treated like FYM & applied if enrichment of FYM is not done.

Fertilizer dose/acre: 40:30:40 kg N:P:K per hectare.

Basal fertilizer: 20:30:40 kg N:P:K, 50% N and 100% P_2O_5 & K_2O (43 kg Urea, 74 kg CAN, 95 kg Ammonium Sulphate + 187 kg Single Super Phosphate + 66 kg Murate of Potash) should be applied to the bed. Mix thoroughly and make sure the bed is levelled.

Laying of drip line: Lay inline drip laterals with 30 cm emitter spacing and 2 L/hr discharge over the beds at the center. Flush drip laterals and check for leakages and blocks.

Polyethylene mulching: Cover the bed with reflective plastic mulch (black–silver), 30-micron thickness and 4 ft width over each bed with the silver side on the top. This helps in water saving, reducing evaporation losses, maintaining soil temperature, reducing weeds and repelling pests.

Transplanting: Make 10 cm diameter holes for seedling transplanting in the plastic mulch on the center of the bed with a spacing of 50 cm between holes. Transplant 14-16 days old seedlings at the center of the hole, transplanted seedlings do not touch the plastic mulch as it may lead to desiccation and death of seedlings especially during hot afternoons.

Irrigation: To ensure constant growth, early irrigation must be done evenly. In order to guarantee a proper fruit set during flowering and to avoid fruit breaking during ripening, less water should be applied to the plants. After the fruits are set, the irrigation flow should be decreased, but watering frequency should be kept constant. One week before harvest, stop watering the fruit to improve its TSS (sweetness). Irregular and over irrigation have must be avoided to prevent fruit cracking.

Fertigation

- **16 DAT to 25 DAT**: 2 kg 19 all+1.5 kg MAP+1.5kg urea fertigation once in 3 days totally 4 fertigation.
- **28 DAT to 43 DAT**: 3 kg 19 all+1.5 kg SOP+1.5kg *$CaNO_3$ fertigation once in 3 days totally 6 fertigation.
- **46 DAT to 60 DAT**: 4 kg 19 all+2 kg SOP+1.5 kg KNo3 fertigation once in 3 days totally 6-7 fertigation.

MAP=mono ammonium phosphate; SOP=sulphate of potash

*inject $CaNo_3$ seperately; Potassium nitrate should be given during fruit growth and maturity phase to improve sweetness.

Intercultural operation: After ten to fifteen days of transplanting, the crop has to have one manual weeding around the plants. The remaining half of the suggested nitrogen dosage should be administered as a top dressing around the plant 30 days after transplanting. De-leafing old and diseased leaves 40–45 days after transplanting helps prevent disease progression and increases photosynthetic efficiency. At this point, vine guiding can also be used to support intercultural activities. When fruits reach maturity, care should be taken to keep them off the wet ground surface by placing some support underneath to prevent rotting.

Foliar nutrition: Fruit set will be improved 40 days after transplanting during the blooming period by applying 2-3 sprays of borax @ 25 ppm (25 mg / L) or any boron-containing formulation every week together with 19:19:19 @ 3g / L. A micronutrient formulation developed by ICAR-IIHR, Arka Vegetable Special @ 3 g/L, is sprayed on the crop 50-55 days after transplanting at the fruit set and development stages to improve fruit quality and yield.

Insect pest management: Control measures need to be taken to manage aphids, mites, leaf miner, thrips, pumpkin beetle, fruit borer and fruit fly.

Disease management: Control measures need to be taken to manage damping-off of seedlings, downy mildew, powdery mildew, Gummy stem blight, anthracnose and fusarium wilt.

10

Cucumber

Vivek Hegde[1], M. Pitchaimuthu[1] and S. Darshan[2]

[1]Division of Vegetable Crops, ICAR-Indian Institute of Horticultural Research Hessaraghatta, Bengaluru - 560 089

[2]HM Clause India Pvt Ltd. Rue Louis Saillant, 26 800 Portes-lès-Valence, France

Cucumber, both slicing and pickling types, is one of the most popular vegetable crops of the family Cucurbitaceae. Like other cucurbitaceous crops, the cucumber also requires a warm season for commercial cultivation and has little or no tolerance to frost. Temperatures above 20°C are optimal for cucumber growth and development. Currently, its cultivars are generally classified as either pickling or slicing types, each with its own set of quality attributes. Pickling types typically have more prominent warts (tubercles) and lighter-coloured skin. They are harvested more frequently at the immature stage, when their length and diameter (LD) ratios are lower. The LD ratio for pickles typically falls within the range of 2.8 to 3.2. On the other hand, slicing cucumbers are harvested and commonly sold in lengths ranging from 12 to 25 cm, with diameters ranging from 1.5 to nearly 7.5 cm. Most slicing cucumbers have cylindrical-shaped fruit with blocky or slightly rounded ends that taper slightly from the stem to the blossom end. Generally, slicers have an LD ratio greater than 4.0, a smooth surface, and thicker skin compared to pickling types. This difference is accentuated because slicers are harvested at later stages of maturity than pickling cucumbers.

Gherkin is a term mostly used to refer to a pickled cucumber. Both slicing and pickling cucumbers belong to the different cultivar groups (*Cucumis sativus* var. *sativus*) of the same species (*Cucumis sativus*). The pickling cucumber is usually fresh-processed (quick-pasteurization) or brined and then processed. Processed cucumbers are available in several sizes and different forms (whole half, strips, chips, cubed and diced) and flavours (dill, garlic, sweet and sour). Recently, India has become the centre of the finest quality gherkin cultivation, processing and exporters to the ever-growing requirement of the world. In India, gherkin cultivation is mostly concentrated in the southern states concentrated in Karnataka, Tamil Nadu and Andhra Pradesh. About 60 per cent of the total gherkin produced in Karnataka followed by 20 per cent each in Tamil Nadu and Andhra Pradesh (Anonymous, 2005). The ideal soil type and desirable temperatures of these areas not less than 15°C and not more than 35°C are prevalent throughout the year, making these regions ideal conditions to cultivate three crops of gherkins. Currently, there are more than one lakh small farmers engaged in the cultivation of gherkins in India.

India has become the largest exporter of cucumbers and pickling cucumbers in the world. During the year 2020-21, India has exported about 223,515.51 MT of cucumbers including both slicing and pickling types to the world for the worth of Rs. 1,651.83 crores/ 223.05 USD Millions with major destinations being the United States of America, Russia, France, Germany and Belgium (https://apeda.gov.in/apedawebsite/SubHead_Products/Cucumber_and_Gherkins.htm). Pickled cucumber is a major dietary constituent to USA, Russia and many European countries, however, it is not palatable with Indian taste. Hence, little or no demand in Indian domestic market except in some star hotels and the entire volume of gherkin produced in India is exported (Acharya, 2006).

Cucumber domesticated in India and wild species *C. hardwickii*, is growing in natural habitats of the Himalayas foothills and it is believed to be the progenitor of cultivated cucumber. Leaves of the cucumber have 3-5 lobes and are obscurely trigonous or cylindric. Cucumber is cultivated in tropical and subtropical regions of India. However, *C. sativus* var. *sikkimensis* is adapted to temperate and humid climates and has 7-9 lobed leaves, an ovary often penta placentiferous, ovoid to oblong fruits. *C. hardwickii* is adapted to cold climates and distributed in natural habitats in the western foothills of the Himalayas and western ghats.

India Vegetable Seed Industry

India is the second-largest vegetable producer in the world after China. The market size of the Indian vegetable seed industry is estimated at USD 700.99 million in 2024 and by 2030 it is expected to reach USD 970 million. The fastest growth of the industry is due to an increase in domestic consumption and exports of fresh and processed vegetables. The availability of improved hybrids having the traits like high, yield disease resistance, wider adaptability and increased shelf life in different vegetable crops increased in the cultivation area under hybrid seeds. Hence, there is an increase in the demand for vegetable hybrid seeds (www.mordorintelligence.com/industry-reports/vegetable-seed-market-india).

India is the world's second-largest producer of vegetables after China. The demand for high-yielding improved hybrids and varieties along with resistance to biotic and abiotic stresses is increasing in India due to the increasing population, decreasing arable land in addition 80% of Indian farmers are small and marginal, the use of resistant hybrids and varieties decrease the input costs. The major vegetable crops grown using hybrid seeds are tomato, chilli, brinjal, okra and cucurbits, and also the varietal development in in these crops are high mostly being released from the public sector. Open-pollinated varieties are more common in the vegetable crops like bulb and tuber crops like tapioca, yams and sweet potato along with leafy vegetables. In India, the share of hybrids and open-pollinated varieties are about 68.7% and 31.3%, respectively of the vegetable seed market. The hybrid and open-pollinated varieties seed market are growing at almost similar rates, i.e., 4.6% and 4.4% respectively in India (www.mordorintelligence.com/industry-reports/vegetable-seed-market-india).

Protected cultivation of cucumber is now expanding in the country and is improving productivity, yields and quality. The incorporation of gynoecy and parthenocarpy are the key traits used in the development of commercial hybrids particularly in the protected cultivation segment. The quality of the products has improved significantly, in addition to helping to produce seedless cucumbers through parthenocarpy. Seed companies like Rik Zwan and Enza Zaden are mostly dominating this segment. In the open field segment, companies like East West Seeds, Rik Zwan, BASF, Chia Tai, H M Clause, Taki and Sakata are the major market leaders. Among Indian companies, Namdhari and VNR are also having an appreciable market share. Gherkins or pickling cucumbers are mainly processed and exported. BASF and Rik Zwan are the leading companies in this segment.

Breeding Objectives of Slicing Cucumbers/Traits for Developing Varieties/ Hybrids

- Early fruiting/ formation of first female flower on lower nodes.
- High female-to-male sex ratio or gynoecious.
- Gynoecious and parthenocarpic for protected cultivation.
- Attractive green or dark green fruits with smooth surfaces and without prominent spines or prickles.
- Uniform long cylindrical shape without crookneck.
- Desirable flesh colour (white, creamy white, green, orange).
- Fruits free from carpel separation.

- Fruits without placental cavity/ hollow spots.
- Fruits free from bitterness.
- Fewer seeds at edible maturity.
- High yield.
- Resistance to biotic stresses like powdery mildew, downy mildew, anthracnose, cucumber mosaic virus, insect pests (melon flies, fruit borer, thrips, aphids) and abiotic stresses (Chadha and Lal, 1993).

Breeding Objectives of Pickling Cucumbers/Traits for Developing Varieties/ Hybrids

- Fruits should have smaller length/diameter ratios.
- Bearing of first female flower on lower nodes.
- More flesh thickness.
- Desirable skin and flesh colour, should have uniform lighter colour.
- Fruits free from bitterness (low or free from cucurbitacin content).
- Cultivar should produce uniform fruits (smaller size more value).
- Less seed or seedlessness (parthenocarpy)
- Gynoecious and parthenocarpic types are mostly preferred.
- Resistance of fruit to balloon bloating in brine tanks, fruits should be free from placental hollowness.
- High yielder.
- Resistance/tolerance to biotic stresses like powdery mildew, downy mildew, anthracnose, cucumber mosaic virus, insect pests (melon flies, fruit borer, thrips, aphids) and abiotic stresses.

Varietal Protection: DUS Guidelines

To encourage the development of new crop varieties to accelerate agricultural production, it has been considered necessary to recognize and protect the rights of the farmers and plant breeders. Such protection is likely to allow the growth of the seed industry which will ensure that the farmers have access to high-quality seeds and planting material.

Distinctiveness, Uniformity and Stability (DUS) tests are used to determine whether a newly bred variety differs from existing varieties, whether the characteristics used to determine distinctiveness are uniformly expressed and whether these characteristics will not change over subsequent generations. DUS descriptors to protect the newly bred variety/ hybrid in cucumber are given in Table 1.

Table 1: Table of characteristics used to conduct DUS testing in cucumber

S.No.	Characteristics	Method of data record	Score	Stage of observation (Days after sowing)
1.	Plant: growth habit	Determinate	1	30
		Intermediate	2	
		Indeterminate	3	
2.	Plant: main vine length	Short (<1.25m)	3	40
		Intermediate (1.25 - 2.0m)	5	
		Long (>2.0m)	7	
3.	Leaf blade: Orientation	Erect	1	30
		Horizontal	2	
		Drooping	3	
4.	Leaf blade: length	Short (<14cm)	3	30
		Medium (14-20cm)	5	
		Long (>20cm)	7	
5.	Leaf blade: shape of apex of terminal lobe	Acute	1	30
		Obtuse	2	
		Rounded	3	
6.	Leaf blade: intensity of green colour	Low	3	30
		Medium	5	
		High	7	
7.	Leaf blade: blistering	Absent	1	30
		Present	9	
8.	Leaf blade: undulation of margin	Absent	1	30
		Present	9	
9.	Leaf blade: dentation of margin	Weak	3	30
		Medium	5	
		Strong	7	
10.	Stem: pubescence	Absent	1	30
		Present	9	
11.	Stem: shape	Angular	1	30
		Rounded	2	
12.	Tendril	Single	1	30
		Branched	3	

13.	Appearance of first pistillate flower in 50% plant	Early <40 days	3	30
		Medium 40-55 days	5	
		Late >55 days	7	
14.	Plant: sex expression	Monoecious	1	30
		Gynoecious	2	
15.	Plant: number of pistillate flowers per node	Solitary	1	30
		Multipistillate	3	
16.	Ovary: colour of vestiture	White	1	30
		Black	2	
17.	Parthenocarpy	Absent	1	40
		Present	9	
18.	Fruit: length	Short (<15cm)	3	40
		Medium (15-25cm)	5	
		Long (>25cm)	7	
19.	Fruit: diameter	Small (<3cm)	3	40
		Medium (3-5 cm)	5	
		Large (>5)	7	
20.	Fruit: shape	Elongate	1	40
		Oblong	2	
		Cylindrical	3	
		Oval	4	
21.	Fruit: shape at peduncle end	Flat	1	40
		Acute	2	
		Obtuse	3	
22.	Fruit: shape at blossom end	Acute	1	40
		Obtuse	2	
		Rounded	3	
		Flattened	4	
23.	Fruit: colour of skin at market stage	Creamy white	1	40
		Yellow (YWG- 158B)	2	
		Light green (GG- 142C)	3	
		Dark green (YGG-146A)	4	

24.	Pulp texture	Crispy	1	40
		Mealy	3	
25.	Fruit: ribs	Absent	1	40
		Present	9	
26.	Fruit: sutures	Absent	1	40
		Present	9	
27.	Fruit: creasing	Absent	1	40
		Present	9	
28.	Fruit: type of vestibular	Hairy	3	40
		Non-hairy	5	
		Prickles	7	
29.	Fruit: density of vestiture	Sparse	3	40
		Medium	5	
		Dense	7	
30.	Fruit: warts	Absent	1	40
		Present	9	
31.	Fruit: stripes	Absent	1	40
		Present	9	
32.	Fruit: length of peduncle	Short (<2cm)	3	40
		Medium (2-3.0cm)	5	
		Long (>3.0cm)	7	
33.	Fruit: colour of skin at ripening stage	White	1	50
		Yellow (160-A)	2	
		Orange	3	
		Brown (175-B)	4	
34.	Seed: size	Small (<1.00cm)	3	50
		Medium (1.00 1.20cm)	5	
		Large (>1.20cm)	7	
35.	Seediness (no. of seeds/ fruit)	Low (75-100)	3	50
		Medium (100-150)	5	
		High (>150)	7	

Fig. 1: Variation observed in stem-end fruit shape and blossom-end fruit shape

Fig. 2: Variation observed in fruit skin colour and fruit spine colour

Fig. 3: Variation observed in colour of strips and presence of placental cavity

Botany and Floral Biology

The plant is creeping or climbing annual, tendril bearing, sparingly branched and scabrous. The stem is obtusely quinquangular, robust, green to yellowish green, densely hispidulous and 50-750 cm long. Tendrils are occurring beside the leaves produced at the node, 10-30 cm long, simple, robust, peduncled, yellowish green in colour and hispid. Leaves are long petioled, distichous, in outline broadly ovate, simple and angular. The base is deeply cordate, apex is acuminate acute, distinctly 3-7 lobed, with unequal glandular teeth, hispid along the margins, light or dark green, beneath pale green or yellowish green, dull on both surfaces, densely patently hispid, very scabrid, 5-7 nerved, the lowest main nerves branched on the outer side, lateral nerves white, transculent, 7-18 cm long and 7-15 cm wide. Lobes are triangular, acuminate, acute and dentate. Petiole is robust, bractate, furrowed on the upper side, clothed with numerous, very stiff hairs, yellowish green and 3-20 cm long. Cucumber is usually monoecious, flowers are axillary, 3-4 cm diameter, shortly stalked, densely hispid, for the greater part male, female flowers are solitary and male ones in fascicles of 3-7. A stalk of male flowers is thin, lengthening after anthesis, clothed with patent hairs with a much swollen and ovoid base, 0.5 – 1.0 cm long and thick. The receptacle is campanulate or tubular, 0.8-1.0 cm high and densely tomentose. Calyx is segmented on the margin of the receptacle, 5-partite more than halfway down, with widely patent segments, paler on the outside, with numerous longitudinal ribs or nerves, which are clothed with scattered long, stiff hairs, between the nerves densely and very shortly hispidulous, wrinkled on inside, densely clothed with minute gland-tipped hairs and 2-3 cm diameter.

Segments are ovate-oblong, lanceolate or obovate, acute, veined, 1.5-2 cm long and 1 - 1.5 cm wide. Male flowers are broadly ovoid, yellowish white, 0.5-0.7 cm long columns of anthers, which are supported by 2-3 very short, free and white filaments. Female flowers are with an ellipsoid or fusiform, pruinose ovary, clothed with scattered hairs with swollen, ovoid or wartlike bases. Style is simple and stigmas 3-5 and thick. Fruits are pendulous having robust stalks, variable in size and shape, usually oblong, lanceolate or cylindrical, small to large, strait or curved, green or white, pruinose with 5-10 more or less conspicuous longitudinal, white or yellowish-white streaks especially towards the apex and often with scattered spinous tubercles and warts, sordidly yellow or sordidly orange coloured and glabrous when ripe, many-seeded, 10-30 cm long and 3-15cm diameter.

The flesh is white, pale green or orange. Fruit-stalk is much lengthened, robust, 2-5 cm long and hispid. Fruits of hermaphroditic flowers, i.e. of perigynous flowers are more round than fruits of epigynous flowers. Staminate, pistillate & hermaphrodite flowers can easily be distinguished from each other visually. Their pedicels can recognize the staminate flowers. Since hermaphroditic flowers start afternoon for anthesis as well as pistil, their column at the top of the ovary is enlarged in comparison with pistillate flowers and can therefore, easily be distinguished from them. Seeds are flat, ovate-oblong, and rather acute at ends, sharply margined, 0.8-1 cm long and 0.3-0.5 cm broad. The whole developmental process is divided into 8 stages from the bud stage to the stage when the flower is detached from the pedicel end of the fruit.

Anthesis occur from 5.30 to 7.00 am and dehiscence occurs around 4.30 to 5.00 am. Pollen fertility is up to 14 hours and receptivity of the stigma is 12hrs before and 6hrs after flower opening. Pollen fertility is considerable up to noon and is greatly reduced by afternoon (2 pm) and by the evening fertility is found negligible. Receptivity stigma is of very short and pollination is carried out within two hours after anthesis. Rise in temperature causes early drying of stigmatic secretion and rapidly loses the receptivity. Different floral abnormalities like hermaphroditism, mixed inflorescence, dimorphic female flowers, fusion, reduction and increase in floral parts have been observed.

Fig.4. Female and male flowers of cucumber **(A & B)**; longitudinal section of male and female flowers **(C, D)**.

Genetics of Sex Expression

There are six sex types reported in cucumber.

1. Monoecious: Both male (staminate) and female (pistillate) flowers in the same plant.
2. Gynoecious: A plant produces only pistillate flowers.

3. Androecious: A plant produces staminate flowers only.
4. Hermaphroditic: A plant produces only perfect flowers.
5. Andromonoecious: Both staminate and perfect flowers are produced in the same plant
6. Trimonoecious: staminate, perfect and pistillate flowers are produced in the same plant.

In cucumber, the sex differentiation controlled by multiple genes, the primary effect genes being Acr/acr (F/f), M/m and A/a (Tatlioglu, 1993) and the basic sex types are decided by the interactions between these loci. These three genes, Acr, M, and A, an ACC oxidase gene encoding 1-aminocyclopropane-1-carboxylic acid synthetase that converts ACC to ethylene *(CsACO2)*, and a transcription factor G. The main function of *CsACS2* is to inhibit stamen development; *CsACS11* stimulates carpel development; *CsWIP1* is inhibited by *CsACS11* as a carpel inhibitor; and *CsWIP1* can also inhibit the expression of *CsACS2* and *CsACO2* (Pawelkowicz *et al.*, 2019). In genetic and regulatory studies on sex differentiation in cucumber, the master effect genes Acr/acr (F/f), M/m and A/a have been reported the most and are key genes in determining the direction of sex differentiation. The 'Acr or F' gene is partially dominant and controls the development of female flower and determines the level femaleness of the plant; 'Ff' is a strong female plant, 'FF' is a full female plant; the stamen development is suppressed by 'M' gene, while the bisexual flower development is invisibly controlled by 'm' gene; the carpel development is promoted by 'A' gene, while the 'a' gene controls invisibly. The combination of these three sex-determining genes determines the sex type of cucumber. All female plants are produced by F_M_A/a; F_mm A/a is forms hermaphroditic plant; ffM_A_ is gives dioecious plant; male plants by ffmmA_; strong male plants produced by ffM_aa; and ffmmaa is an all-male plant, where F has a dominant epistatic effect on A (Table 1). Environment factors like nutrition, day length, light intensity and stress factors affect sex expression considerably.

Table 2. Genetics of sex expression in cucumber

	Acr (F)	**acr (f)**	
	A or a	A	a
M	Gynoecious	Monoecious	Androecious
m	Hermaphrodite	Andromonoecious	Androecious

Production Technology

Proper selection of soil and climate are very important for seed yield and quality of cucumber.

Soil type: Sandy and sandy loam soils, rich in organic matter with good drainage and a pH range from 5.5-6.8 are the best. Poorly drained soils should be avoided.

Climate: Cucumber requires a moderately warm temperature. It does not withstand frost, minimum temperature should not be lower than 18°C. Growth and development are favoured by temperatures above 20°C. Day and night temperatures of 20-25 °C and 18-20 °C, respectively are preferred for optimum growth. A relative humidity of 70-80% is ideal for the better growth.

Site Selection

A minimum of three years of crop rotation away from the crops belonging to the Cucurbitaceae family is to be followed to prevent the residual presence of disease pathogens and insect pests. The field needs to be fully cleared of crop debris before the field preparation. To ensure there are no potential pests and diseases, all crop residues must be removed or incorporated well into the soil for decomposition. The production site needs to be free of weeds in the surroundings of the plot or field and at the entrance of protected areas, weeds act as a natural host for the number of pests and diseases. Seed production in the open field will follow commercial production procedures. Cucumber seed production is taken in a protected culture within a closed and restricted greenhouse, tunnel or netted cage to avoid the entry of insect vectors. Protected structures need to remain closed at all times and checked periodically for any damage within the structure. If damage is noticed it must be identified immediately and necessary action should be taken. The mesh size of the protected structure should be determined by local regional insect and disease pressure. Mesh size recommendations are 841 microns to control the entry of cucumber beetle, 297 microns mesh for aphids and whiteflies, similarly, 210 mesh microns for thrips.

Spacing: maintain a spacing of 1.5 m between the rows and 60 cm between the plants.

Seed rate: Generally, 400 g/ha, 320 g female and 80 g male parent seeds are required per hectare area.

Nursery raising: Pro-trays are filled with microbial-enriched cocopeat and raised in a protected structure. Seedlings can be drenched with Bavistin (2 g/l of water) or captan (3 g/l of water) on the 10th day after sowing as a precautionary measure against damping off. Seedlings will be ready (with one true leaf) for transplanting by 12-14 days after sowing (Fig. 2).

Fig. 5: Nursery raising technique for the production of quality seedlings; **(A)** Fermented cocopeat for seed sowing; **(B)** Sowing of seeds in portrays; -562109 Covering of portrays after seed sowing with black polythene to enhance the germination; **(D)** Germinating cucumber seeds; **(E)** 14 days old cucumber seedlings which are ready for transplanting in the main field; **(F)** Transplanted seedling in the main field.

Land preparation: Bring soil to a fine tilth by 2-3 ploughing and harrowing. Make a raised bed of 15-20 cm in height, 60 cm width and convenient length depending on the drip system.

Enrichment of FYM: Before application of FYM to the main field as a basal dose, sieved 100 kg of moist FYM is enriched with 1kg each of *Trichoderma viridi*, *Trichoderma harzianum*, *Paecilomyces lilacinus*, phosphorus solubilizing bacteria and Multi-K. Cover the treated FYM with polythene covers for 15 days under shade and would be mixed with 9 tons of well-decomposed FYM before application. FYM must not keep open in direct sunlight for long periods to protect the microorganisms and other beneficial agents like earthworms. This technique will double the NPK content more than the normal FYM and help to control the important pests and diseases (Fig. 3).

Fig. 6: Enrichment of FYM; **(A)** Selection of well-decomposed, moist FYM; **(B)** FYM is enrichment with *Trichoderma viridi, Trichoderma harzianum, Paecilomyces lilacinus*, phosphorus solubilizing bacteria and Multi-K (1kg each/100kg FYM); **(C)** Mixing of beneficial micro-organisms with FYM; **(D)** Covering of treated FYM with polythene covers for 15 days under the shade for multiplication of micro-organisms.

FYM application: apply 20-25 t/ha of enriched FYM at the time of land preparation as a basal dose.

Neem cake application: recommended to apply neem cake @ 250 kg/ha before final bed preparation as a basal dose

Fertilizer dose: a recommended dose of 60:50:80kg/ha N:P:K applied in three split doses during land preparation, vegetative and flower initiation stages.

Transplanting: A diameter of 10cm holes are made in the centre of the plastic mulch at 60 cm spacing. 12-14 days old seedlings are transplanted at the centre of the hole. Care must be taken while transplanting that the seedlings should not touch the plastic mulch as it may lead to the desiccation and death of the seedlings.

Irrigation: Irrigate the field two days before planting. Proper soil moisture should be maintained and over-irrigation should be avoided. Depending upon the soil and weather conditions, irrigate the crop once in 4-5 days.

Intercultural operation: The crop should be kept weed free by giving shallow cultivation and manual weeding. Top-dress the crop with the remaining half dose of recommended nitrogen at 30-35 days after transplanting. Remove cotyledonary leaves 30 days after transplanting to help in checking disease spread. Vine guiding/ training may also be done when the plants start veining (35-40 days after sowing) for facilitating intercultural operations. The plants should be provided suitable support made of iron angles, bamboo or wooden poles particularly in the rainy season to check against the rotting of fruits.

Foliar nutrition: About 30-35 days after transplanting during flowering, 2-3 sprays at weekly intervals of borax @ 25ppm (25mg/L) along with 19:19:19 @ 4g/L improves fruit set. At the fruit set and development stage, about 50 - 55 days after transplanting, spray of a micronutrient formulation developed by ICAR - IIHR, Arka Vegetable Special @ 4g/L at 15 days intervals helps in improving the yield and quality of the crop.

Management of major pests and diseases: Preventive and/or curative insect pest and disease control measures must be followed.

Table 3. Major diseases of cucumber and its management

Diseases	Control measures
Damping-off of seedlings	Drench with Bavistin (2g/L) or captan (3g/L) 10 days after sowing.
Downy mildew	An alternate spray of Ridomil Gold (2.0 gm/L), Cabriotop (1.0 -1.5gm/L), Curzate (2.0 gm/L), Lurit (1.0 gm/L), Aliete (1.0 gm/L) or Equation Pro (1.0 ml/L)
Powdery mildew	An alternate spray of Score (1.5 ml/L), Luna exp (1.0 ml/L), Cabriotop (1.0 - 1.5 gm/L), Nativo (1.0 ml/L), Roko (2.0 g/L) + Biojodi (5.0 g/L) or Sprint (2g/L) + Biojodi (5g/L)
Fusarium wilt	Carbendazim (1.5-2.0 gm/L) as drenching, Folicure/ Score (1.0 ml/L) as spray

Downy mildew

Powdery mildew

Viral disease

Table 4. Major insect pests of cucumber and its management

Pests	Control measures
Aphids	Confidor super (0.3 ml/L), Pegasus (0.3 ml/L)
Leaf miner	Triazophos (2.0 ml/L), Triazophos + Deltamethrin (Jadu, scoop, punch) (2.0 ml/L), Abacin (1.0 ml/L)
Thrips	Regent (1.3 ml/L), Spinosad (0.5 ml/L)
Pumpkin beetle	Acephate (lancer) (2.0 gm/L), Marshal (2.0 gm/L)
Cucumber moth	Kingdoxa (Indoxacarb 14.5% SC) @ 0.8 ml/L
Fruit fly	• Erecting Cue lure or Barrix trap 6-8/ acre. • Spray Neem soap (3-5ml) at 4 days intervals. • In 10 litre of water add 1kg jaggery + 20 ml Deltamethrin and sprinkle on the plants using broom/ brush.

Fruit fly

Cucumber moth

Pumpkin beetle

Training and Pruning Operations for Hybrid Seed Production in Cucumber

1. Single Stem
   - Using a suitable system, train the vine vertically.
   - On the lower part of the plant up to the 5th node remove branches and young fruits.
   - On 8-10 nodes onwards start pollinating the female flower and pollination operation continued up to the 15th - 16th nodes.
   - Setting of 5-6 fruits on the main stem are allowed.
   - All the female flowers are removed after the 16th node.
2. Branching Type
   - Train the vines vertically on a suitable support
   - On the lower part of the vine up to 30 cm remove all branches and young fruits.
   - Start pollination of the female flower produced on the branch emerging from the 6th node onwards.
   - After the pollination and successful fruit set pinch off the tip of each branch.

- Pollinate female flowers is done on the main stem of 9-10th nodes onwards
- About 6-7 fruits/plant are allowed to get optimum seed yield and good quality seeds
- Fruit set lower than 30 cm above the ground should not be allowed, since the fruits which are extra early contain less seed as well as tend to touch the ground and rot. similarly poor quality of seeds are observed on the fruit setting occurs on extremely upper nodes of the vine.

Pollination Operation

- A day prior to anthesis bag or tie both the female and male flowers on both female and male lines, respectively.
- On the next day morning collect male flowers.
- Start pollination in the morning at 7.0 am and complete the operation latest by 10.30 am.
- Pollination may continue for a period of 10-15 days.
- Pollination operation requires 20-25/ labourers/acre/day, depending upon the plant population.

Fig. 7: Pollination in cucumber; male and female flowers ready to open next day **(A&B)**; bagging of flowers a day before anthesis **(C)**; collection of male flowers for pollination **(D)**; pollination of bagged female flower **(E)**; Bagging of pollinated flower **(F)**.

Rouging

For removal of off-types to produce quality seeds, seed crop is to be monitored at various crop growth stages. To avoid natural cross-pollination, first rouging should be carried out at the vegetative stage before flowering. Second should be done during the fruit set and complete fruit development stages based on variability observed in fruit size, colour, shape, colour of ripened fruit, presence of spines, spines colour etc. During crop inspections, observations should be done for Bacterial fruit blotch (BFB), Cucumber green mote mosaic virus (CGMM), Squash mosaic virus (SqMV), Kyni green motte mosaic virus (KGMMV), Melon necrotic spot virus (MNSV), Fusarium wilt, Gummy stem blight (GSB), Fusarium crown and foot rot and *Macrophomina phaseolina*.

Suitable control measures should be taken if observed above-mentioned diseases. Suspected plants are immediately removed and destroyed if observed in the seed crop. Affected plants are uprooted and collected in a large plastic bag. Then the eliminated plants will be burned or burned in a designated area which should be away from the production area.

Methods of Hybrid Seed Production

Artificial/ hand pollination

- Emasculation is not necessary in cucumber since the plants are monoecious and gynoecious.
- The female flowers of the seed-parent (female parent) can be artificially pollinated with pollen collected from the male flowers of the pollen parent.
- A large number of seeds are produced by a single pollination and therefore, the cost of a hybrid production in cucumber is comparatively low.

Removal of Male Buds and Use of Insect Pollinators

- The male and female are grown in the ratio of 4:1 in alternate rows.
- Regularly pinch off all the male flowers a day before anthesis from female lines.
- Honeybees and other insects (voluntary) serve as pollinating agents.
- The male lines are uprooted and destroyed after the pollination period is over to avoid the possibility of the mechanical mixture during the harvest and seed extraction.
- The fruits set on female lines are harvested for seed extraction since this is a hybrid.

Use of Gynoecious Sex Form

- The gynoecious sex form has been commercially exploited in hybrid seed production.
- The gynoecious sex form in cucumber has been commercially exploited in hybrid seed production.
- Female (homozygous gynoecious line) and male (Gynoecious, Hermaphrodite, monoecious) rows are planted in 4:1 ratio.
- Only female flowers are produced in the female/ gynoecious lines and natural pollination is done by the insects (honeybees).
- At the boundary of the seed production plot a sufficient population of honeybee colonies has to be kept to ensure good fruit and seed recovery.
- Hybrid seeds are collected from the female/ gynoecious plants.
- The parental lines i.e. The female lines i.e. gynoecious lines are reproduced and maintained by inducing the staminate flower through the sprays of 200-500 ppm silver nitrate or silver thiosulphate at two to four true leaf stages and then selfing whereas the male parent is maintained by selfing.

Hybrid seed production of cucumber is subject to two phases in this method. Breeding of a female or gynoecious line with a strong genetic female tendency is the first stage. The second stage is hybrid seed production and multiplication of female lines. Hybrid seed production and multiplication of female lines are part of the second stage. The hybrid seed is produced in an isolated field in which generally four rows of female (gynoecious) or seed-parent lines are alternated with one row of pollinator line (gynoecious, hermaphrodite or monoecious). In the cucumber hybrid-seed production field, it is necessary to keep at least four beehives per hectare to ensure effective cross-pollination to get the maximum seed yield. The hybrid seed is collected from female lines. To improve the pollination and yield, it is necessary to blend 10% seed of monoecious cultivar with gynoecious hybrid seed, it is not needed if the gynoecious hybrid is parthenocarpic.

General Recommendations to Induce Male Flower on Gynoecious Lines & Production of Hybrid Seed

Gynoecious lines normally produce only female flowers. To induce male flowers, gynoecious lines are sprayed with GA_3, silver nitrate or silver thiosulphate. Three applications of GA_3 at 1000 ppm or $GA_{4/7}$ at 50 ppm, at fortnightly intervals starting from when the plants have two leaves will induce male flowers. A single application of silver nitrate solution or silver thiosulfate

(400 - 600 mg/l) when the plants have 2-3 leaves before the first flowers open is more commonly used. Silver thiosulfate at 3mM is very effective and more commonly used. Since it induces male flowers for an extended period, less phytotoxic and does not cause excessive stem elongation or malformed male flowers as observed with the gibberellin application. It is prepared by mixing of 5.95g/l of Sodium thiosulfate and 0.51g/l of silver nitrate. Sodium thiosulfate and silver nitrate are prepared separately using distilled water and always add the $AgNO_3$ solution to the $NaS_2O_3.5H_2O$ solution.

Harvesting, Curing and Seed Extraction

Ideally, the fruit should remain on the plant until it is fully mature (Fig. 5 A&B). Harvest the fully matured fruits 35-50 days after pollination, depending upon the variety. The maturity of the fruits of the seed crop is indicated externally by the development of the ripe rind colour (characteristic of the cultivar), when the seed is mature the fruit stalk adjacent to the fruit withers and mature seeds separate easily from the flesh. Cure the fruits for 5-7 days under shade in a cool and dry place. The mature fruits are handpicked and placed in a fruit crusher and seed extractor or if the seeds are extracted manually, the ripe fruits are cut in half longitudinally and the seeds are scraped and collected in the container.

Seed extraction is done by two methods, natural fermentation and acid treatment.

1. **Natural fermentation**
   - Fruits are sliced in half longitudinally and collected the seeds with surrounding pulp.
   - Allowed to ferment for about 24 hours under room temperature.
   - Seeds are separated from the fermented pulp and washed with clean water.
   - Clean seeds are dried in nylon mesh bags under the sun.
2. **Acid treatment**
   - Seeds are collected with surrounding pulp from the ripened fruits.
   - Commercial hydrochloric acid (3 fluid ounces/ 88.72 ml) or sulfuric acid (1 fluid ounce/ 29.57 ml) was added to and stirred with 11.3 kg (25 pounds) of freshly extracted seeds with pulp.
   - Water is added to the mixture after 15 to 30 minutes and the digested pulp floats to the top while the mature seeds sink to the bottom.
   - Wash the seeds with plenty of clean water to remove trace of acid and the pulp.
   - Dry the seeds under sunlight for 3-4 days in nylon mesh bags.

Fig. 8: Harvesting, curing and seed extraction; Fully mature fruits **(A&B)**; Curing of fruits for under shade in a cool and dry place; manual extraction of seeds **(D&E)**; Fermentation of seeds with surrounding pulp **(F)**; washing of seeds **(G)**; Sun drying of washed seeds **(H)**

Seed Yield

- About 300-350 kg/ha (1:4 male: female plant population, total plant population =19,000 plants/ha)
- The seed yield from a single fruit is approximately 300 - 500 seeds/fruit, depending on the cultivar, crop management and the amount of successful pollination.
- The average seed yield under field conditions of open-pollinated variety is 400 kg/ha and up to 700 kg/ha.
- The seed yield for F_1 hybrids produced in fields with a male: female parental population ratio of 1:4 is 300-350 kg/ha.
- The 1000 seed weight is approximately 25g in smaller fruited cultivars and 33g in the longer fruited cultivars.

Seed Certification Standards

1. Field Inspection

A minimum of three field inspections should be done, the first during the vegetative stage before flowering, the second during the flowering and fruit development stage and the third at the mature fruit stage and before harvesting.

2. Field Standards

Cucumber seed fields should be isolated from other pollinizer or contaminants like fields of other varieties of the same crop, fields of the same variety not conforming to varietal purity and from cucumber wild species *Cucumis hardwickii* which readily crosses with the cucumber. Cucumber is not cross-compatible with any other cucurbitaceous members like *C. melo*, squash, gourds, pumpkins, marrows or watermelons. The recommended isolation distance for cucumber seed production is 1500m for breeder seed, 1000m for foundation seed and 800m for certified seeds.

3. Seed Standards

Table 5: Genetic purity and seed health standards

Factors	Maximum permitted level (%)	
	FS	CS
Open pollinated variety		
Off-type	0.1	0.2
Objectionable weed plant*	None	None
Hybrids		
Off-type in seed parent	0.01	0.05
Off-type in pollen parent	None	0.05

*Objectionable weed - *Cucumis hardwickii* Royle.

Table 6: Seed standards

Factors	Permitted level (%)	
	FS	CS
Pure seed (minimum)	98	98
Inert matter (maximum)	2	2
Other crop seeds (maximum)	None	None
Other objectional varieties	5/kg	10/kg
Weed seed (maximum)	None	None
Objectionable Weed seeds (maximum)	None	None
Germination (minimum)	60	60
Moisture for ordinary pack (maximum)	7	7
Moisture for vapour-proof pack (maximum)	6	6

11

French Bean

C. Mahadevaiah, M.V. Dhananjaya, H.S. Yogeesha and T.S. Aghora

ICAR-Indian Institute of Horticultural Research, Hesaraghatta Lake Bengaluru-560089, Karnataka

Introduction

French bean (*Phaseolus vulgaris* L.) is a nutritious protein-rich legume vegetable crop cultivated in all parts of the world except Antarctica. Immature pods are consumed as vegetables, and matured pods are used as grain legumes (Allem, 2013). French bean is cultivated in 1.65 million hectares of area with a production of 26.98 million tonnes and productivity of 16.36 tonnes per hectare for vegetable purposes. As a grain legume, French bean was cultivated in 33.02 million hectares with a production of 28.90 million tonnes and productivity of 874.1 kg/ha during 2019 globally. In India, French bean is grown in 2.56 lakh hectares for vegetable purposes and 126.91 lakh hectares for grain legume purposes with a production of 7.26 lakh tonnes as immature pods and 53.10 lakh tonnes of grain legumes with a productivity of 2.83 t/ha of green pods and 418.4 kg/ha as grain legumes (area, production and productivity of livestock, Food and Agriculture Organization of United Nations, 2020). However, the area, production and productivity statistics of beans in FAO statistics may also include other beans such as mung bean and others (Sharma *et al.*, 2013). French bean is also known by other names such as common bean, kidney bean, dwarf bean, haricot bean, string bean and garden bean. In India, French bean is also known by different names as *Frash bean* and *Rajmah* in Hindi, *Shravanghevada* in Marathi, *Thingalaware* in Kannada and *Fras bean* in Punjab.

Geographic Distribution

French bean is cultivated in a wide range of climatic conditions across the world. The crop is distributed in the latitude range of 30°N to 35°S and an altitude range of 500 to 2000 MSL, even up to an altitude of 3000 MSL and distributed in rainfall zones of 800-1800 mm (Gepts, 1998; Singh *et al.*, 1988).

French bean is a cool season crop, the optimum temperature for good plant growth and development ranges from 17.5°C to 25°C and is better at a mean temperature of 21°C and there are many varieties and germplasm with better adoption to warmer climates. Extreme high temperatures interfere with pod filling, while low temperatures are unfavourable for vegetative growth. A favourable soil temperature required is 18°-24°C.

Floral Biology and Pollination Behaviour

French bean's flower is a typical Papilionaceous floral arrangement comprised of five petals *viz.,* a central wing petal, two wing petals and a keel petal formulated into boat-like structure (OECD, 2016). The keel petal is formulated into a boat-like structure and coiled in two or three turns around the stamens and the stigma promotes cross-pollination. There are 10 stamens, nine form the tubes on the ovary/stigma and another one is positioned above the ovary. French bean is a cleistogamous, and anthesis occurs after pollination. Anther sacs are borne directly adjacent to the stigma, and anther bursts on a day before anthesis and releases pollen onto the stigma ensuring complete self-pollination. Stigma is highly receptive even two days before the anthesis and crossing can be attempted at the flower bud stage two days before the crossing. This is the general practice followed at ICAR-Indian Institute of Horticultural Research, Hesaraghatta Lake Post, Bangalore.

Breeding Objective- Breeding Methods

The major breeding objectives of varietal development programmes aim at the development of varieties in string/snap/haricot bean aimed at the development of varieties for different pod quality parameters combined with resistance to biotic and abiotic stresses. French bean varieties are broadly categorized into two segments *viz.,* bush bean (determinate growth habit) and pole bean (indeterminate growth habit) based on their growth habit of cultivars. Further based on pod quality parameters, cultivars are broadly categorized into round and flat beans, stringed and stringless pods, and dark green/green/light green pods. Generally, varieties developed by public institutions have both flat and round pods with stringed and stringless pods, whereas varieties developed by Private Institutes have varieties with round-shaped stringless pods with dark green, green and light green-coloured pods. Besides, resistance to many fungal (diseases such as rust, anthracnose, angular leaf spot), bacterial (common bean bacterial leaf blight) and viral diseases (Horse Gram Yellow Mosaic disease and bean common mosaic disease) is most desirable.

Major breeding methods adopted in the breeding programmes are Pedigree breeding and backcross breeding methods. Pedigree breeding methods adopted of selection in biparental segregating populations and backcross breeding programmes are preferred methods for the transfer of resistance genes for rust, anthracnose and other diseases. SCAR markers are available for most diseases (http://www.bic.uprm.edu/) and marker-assisted selection is being practiced in bean improvement programmes.

Major centres of bean improvement programmes include CGIAR institute CIAT (The International Center for Tropical Agriculture), Cali, Colombia, which maintains about 37938 *Phaseolus* accessions belonging to 44 taxa originating from 112 countries (https://ciat.cgiar.org/what-we-do/crop-conservation-and-use/bean-diversity/). Besides, USDA is also a major research centre and a total of 17940 accessions of *Phaseolus* spp. and 14075 accessions of *P. vulgaris* are maintained. Besides, many international research organizations in various countries are involved in research and development activities in French bean. In India, ICAR-Indian Institute of Horticultural Research, Hesaraghatta Lake Post, Bangalore, ICAR-Indian Institute of Vegetable Research, Varanasi, ICAR-IARI Regional Station, Katrain, Tamil Nadu Agricultural University Coimbatore and many other varieties are associated with breeding research activities in French bean

Varietal Achievements: Market Segments, Current Industry Priorities Important Companies and, Ruling Hybrids and Reasons

Market Segments

The global market segments in French bean are broadly categorized into pole bean (indeterminate growth habit) and bush beans (determinate growth habit). Pod quality parameters are chief features that determine the market segments or consumer preferences. Based on pod shape/cross-section, French bean cultivars are broadly grouped as round, oval and flat pods and based on pod colour, it is categorized as green, light green and dark green coloured. The presence of a string in pods is categorized as stringed and stringless pods.

Among bush-type cultivars, ICAR-IIHR, Bangalore has developed five varieties. Arka Komal (Selection-9) and Arka Arjun are green-coloured flat pods and string is present in the pods. Whereas Arka Suvidha is oval-shaped green stringless pods. Arka Sharath and Arka Arjun are green-coloured smooth stringless pencil/round-shaped pods with high consumer preferences in the markets. ICAR-Indian Institute of Vegetable Research, Varanasi has developed three varieties such as Kashi Rajahans (dark green round-shaped pods), Karshi Param and Kashi Sampann (flat green pods).

Major private seed industries having significant market share to our best understanding are Ashoka Seeds (NZ, US-2, Super NZ, Indradansh), Bayer

Crop Science (Falguni, Moraleda, Alhama), Syngenta (Serengeti), East-west seeds (Vaishnavi), Sakata seeds (Mridula), Namdhari seeds (NS 620). Major seed industry priorities include capturing the market related to dark green round stringless pod segments with superior pod quality and fewer industries focusing on flat segment categories. Arka Komal (Selection-9) is a still ruling variety with a significant market share in the flat segment as farmers can maintain their own seeds for sowing.

Varietal Protection: Dus Testing Guidelines with Photos of Different States, List of Varieties Already Protected

The Protection of Plant Varieties and Farmers Rights Authority (PPVFR Authority), Government of India is a nodal organization involved in the protection of plant varieties in India. A total of 22 DUS descriptors were developed for French Bean or kidney bean (https://plantauthority.gov.in/sites/default/files/gkidney-bean_1.pdf). The list of protected varieties under PPVFR authority as of 2603.2024 () is given in Table 1.

Table 1. List of protected varieties in PPVFR Authority in Kidney bean.

Sl. No.	Category of variety	Name of Variety	Applicant
1	Extant (Notified)	Utkarsh (IPR 98-5)	Indian Council of Agricultural Research (ICAR)
2	Extant (Notified)	Kashi Puram (IVFB-1)	Indian Council of Agricultural Research (ICAR)
3	Extant	AMBER (IIPR 96-4)	Indian Council of Agricultural Research (ICAR)
4	Extant	Shalimar Rajmash-01	Indian Council of Agricultural Research (ICAR)
5	Extant	Arka Suvidha (IIHR-909)	Indian Council of Agricultural Research (ICAR)
6	Extant	VL Bean-2	Indian Council of Agricultural Research (ICAR)
7	Extant	Varun (ACPR-94040)	Indian Council of Agricultural Research (ICAR)
8	Extant	Arun (IPR 98-3-1)	Indian Council of Agricultural Research (ICAR)
9	Extant (VCK)	ARKA ANOOP	Indian Council of Agricultural Research (ICAR)
10	Extant (VCK)	ARKA BOLD	Indian Council of Agricultural Research (ICAR)
11	Farmer	SAFED RAZMAH	Gazenfer Gaffar
12	Extant (Notified)	Phule Rajmah (GRB-902)	Mahatma Phule Krishi Vidyapeeth

13	Extant (VCK)	Alhama	Monsanto Holdings Pvt Ltd
14	Extant (VCK)	Moraleda	Monsanto Holdings Pvt Ltd
15	Farmer	Safed Jhulu Swant	Gram Panchayat Malla Ghorpatta
16	Farmer	Chamba Rajmash	Bhandal Panchayat
17	Farmer	Safed Swant	Gram Panchayat Suring
18	Farmer	Chitkabra Lal Jhulu Swant	Gram Panchayat Sarmouli
19	Farmer	Kanchan Rajma	Arvind Kumar
20	Farmer	THUL RAZMAH	Ajaz Ahmad Pir

Commercial Practice of Seed Production and Seed Production Areas

i) Soil and Climatic Conditions for Seed Production

French bean is grown over a wide range of well-drained soils and better crop growth is recorded with loamy soil. Clay soils impede the emergence of seed growth leading to uneven or poor stand. The well-aerated soil is ideal for optimum nitrogen fixation and the soil pH range of 5.5 to 6.8 is very ideal for good crop growth. French bean is highly sensitive to acidic and alkaline soils, and hence, it is not recommended to take seed production in the problematic soil.

French bean is a tender, cool-season vegetable crop that cannot tolerate frost. Seeds do not germinate if the temperature goes below 15°C and flowers drop in hot or rainy weather. A mean air temperature of 20°-25°C is optimum for its growth and high pod yield. The extremely high temperature results in partial pod filling and the low temperatures restrict the vegetative growth. Besides, shade drying of seeds is better in February and March months with lesser humidity (30-35%) to attain the optimum moisture content of 8.0 - 9.0% in seeds. Considering all these factors, *rabi* season is ideal or optimum season for seed production in Southern India, October-November sowing provides optimum climatic conditions for crop growth, good pollination and quality seed production, whereas February and March months are ideal for shade drying of seeds to attain the optimum moisture content and goof seed health.

ii) Land Preparation and Nutrition

The deep ploughing of soil for two-three times followed by the harrowing of the field is essential to attain the fine tilth. A recommended FYM enriched with *Trichoderma* and phosphorous solubilizing biofertilizers has to be incorporated into the soil before sowing. Formation of ridges and furrows with a spacing of 60 cm between rows and the application of Neem cakes (250 kg/ha) at the time of ridges and furrows formation is essential.

The requirements of nitrogen, phosphorous and potassium fertilizers depend on soil and agroclimatic zones. Soil health tests are essential for the recommendation of NPK fertilizers. However, the general recommended dose of NPK is 60:50:70 kg/ha. The complete phosphorous fertilizer is recommended to be provided as a basal dose, whereas 50% of nitrogen and Potassium fertilizers are a basal dose and the remaining 50% as topdressing at the time of earthing up operations after the third week of sowing. This is generally followed for bush bean and pole bean generally requires higher dosage of NPK fertilizers considering the longer duration of crop growth and high biomass production. The foliar application of micronutrients such as Arka Vegetable Special (2-3 g/L) and 2-3 g/L of water-soluble potassium (Sulphate of Potash contains 50%K) from pre-flowering to pod filling stages enhances the good pod set and seed filling in many vegetable crops including French bean.

iii) Seed Rate and Spacing

The recommended seed rate for bush and pole bean is 35-40 kg/ha and 10-12 kg/ha respectively. It is recommended to seed treatment with *Rhizobium phaseoli* for quick nodulation and fixation of atmospheric nitrogen. The recommended spacing for bush bean and pole beans is 60 cm × 10 cm and 120-150 cm × 45-50 cm respectively. Though the higher plant population/spacing increases seed yield, but reduces the seed colour intensity and uniformity of French bean. Hence, the recommended seed rate and spacing have to be adopted for the production of quality seed production in French bean.

iv) Integrated insect and Disease Management in French Bean Seed Production Programmes

The major diseases in French bean include bacterial leaf blight disease, bacterial halo blight, bean rust, angular leaf blast, anthracnose, root rot disease and insect pests include white flies, aphids, thrips, mites and pod borer in endemic areas. The integrated insect pest and disease management approaches help harvest the maximum seed yield.

v) Harvesting and Processing

Most of the French bean have indehiscent pods and shattering is not a problem. Fully matured pods have been harvested. Manual threshing of pods is recommended to avoid mechanical mixtures and seed damage. The seeds have to be sundried in the shade during the February-March months or in a polyhouse to avoid seed coat damage and the moisture content in the seeds has to attain 8.0.9.0 per cent. Normally 1500 to 1800 kg of good quality seed can be harvested in one hectare.

Seed Standards

i) Isolation Distance: Gene flow is also common in self-pollinated crops and outcrossing is expected around 0.0074 % if two flowers are apart by 2.5 meters to 0.136% if the distance between two plants is 0.5 meters and natural outcrossing is zero if two varieties are apart with the distance of 3.25 meters. Therefore, 15 15-meter isolation distance is recommended for commercial seed production in French bean depending upon the floral traits and varietal characteristics (De Ron *et al.*, 2015). However, as per the minimum seed certification standards, a minimum of 10 meters and 3.0 meters isolation distance is recommended for Foundation and Certified Seed Production Programmes respectively from other varieties and same varieties not conforming to the specified seed certification standards (Trivedi and Gunasekaran, 2013).

ii) Off-types: As per the minimum seed certification standards, the maximum permissible off-types and seed-transmitted diseases such as *Xanthomonas* leaf blight, anthracnose, *Ascochyta* leaf blight and bean common mosaic disease infected plants should not exceed 0.1 and 0.2% for certified and foundation seed production respectively (Trivedi and Gunasekaran, 2013). The Regular rouging at vegetative, pre-flowering and pod formation stages is required to remove the off-types.

iii) Seed Purity Standards: According to the Indian minimum standards for seed certification, minimum seed purity should be 98%, and maximum permissible inert matter and maximum moisture contents are 2.0 and 75% respectively whereas in the vapour-proof containers, the maximum permissible moisture content is 7.0% in both Certified and Foundation seeds. Other crop seeds should be nil in both Foundation and Certified Seeds, weed seeds should be nil and a maximum of 10 per kg, and other distinguishable varieties should be a maximum of 5 per kg and 10 per kg in Foundation and Certified Seeds respectively (Trivedi and Gunasekaran, 2013).

Seed Processing and Machinery

The seed processing is essentially required to maintain uniformity in seed colour and seed size. removal of seed coat and seed treatment with pesticides and polymers. The seed coat treatment with polymers along with nutrients and plant protectants gives aesthetic value to the seeds and also improves the seed germination and plant vigour. Many equipment such as seed driers, seed sieving machines, gravity separators, seed colour sorters, seed polymer coating machines and other high-end equipment are required for seed processing in the industries.

12

Dolichos Bean

M. Thangam

Division of Vegetable Crops, ICAR-Indian Institute of Horticultural Research Hesaraghatta Lake Post, Bengaluru- 560 089, Karnataka

Introduction

Dolichos bean, [*Lablab purpureus* (L.) Sweet] var. *typicus* (2n=22) also known as Indian bean or *sem* is one of the vital leguminous vegetables grown in India for its tender green pods. Both pole types and bush types are commercially grown. It is predominantly cultivated in Maharashtra, Tamil Nadu, Andhra Pradesh, Uttar Pradesh and North-East India. Protein, minerals, vitamins and fibers are abundant in Dolichos bean. Bush-type vegetable dolichos unlike the pole-type dolichos (which are generally grown during the winter season as they require short day) are thermosensitive and can be grown year-round.

Dolichos is a versatile and valuable leguminous crop, offering multiple benefits. It boasts a high protein content of 3.6% and fiber content of 1.8%. The dry seed is composed of a protein content ranging from 23.0% to 28.0%. The pods are abundant in phenols, with levels ranging from 1.7 to 9.67 mg/100 g, making them a valuable source of antioxidants. Furthermore, dolichos serves as an excellent source of amino acids, particularly lysine, along with vitamins such as A, C, and riboflavin, and essential minerals (Ca, Fe, Mg, S, Na and P). The seeds of dolichos contain a flavonoid, kievitone, recognized for its potential in combating breast cancer. Additionally, the presence of tyrosinase in the seeds indicates promising possibilities for treating hypertension in humans. Dolichos beans are utilized for various medicinal purposes, acting as stomachic, anthelmintic, diuretic, aphrodisiac, antispasmodic, digestive, febrifuge, and carminative agents. Furthermore, they are rich in bioactive compounds that offer significant benefits in treating specific ailments such as liver problems, diabetes and tumors.

Seed Requirement and Market Size

Crop	Area (Lakh. ha.)	Seed rate (kg/ha.)	Seed Requirement (ton.)	Actual requirement @30% SRR	Seed price (Rs. /kg.)	Total seed value (in Crores.)
Dolichos bean	0.9	30	2700	810	300	24.30

Description

The Indian bean is a tropical, herbaceous dicotyledon with a thermosensitive flowering plant exhibiting annual growth. It features robust twining vines with woody stems and large, heart-shaped trifoliate leaves comprising broad ovate-rhomboid leaflets of 7 to 15 cm in length and 4.5 to 15 cm in width. The upper side of the leaf is smooth, while the lower side is hairy. The plant possesses a tap root system, with the main root giving rise to numerous lateral roots distributed evenly along its axis. Its flowers are bisexual, hypogynous and pentamerous, displaying a range of colors from white to pink to purple. The inflorescence is axillary, erect, and arranged in racemes.

Floral Biology and Pollination Behavior

Floral Biology: Flowering typically occurs between September and February, with numerous small purple or white flowers arranged in elongated racemes. The stigma is wet, papillate and the style is solid. Flowers are shortly pedicelled, violet or white, in long-stalked many-flowered, axillary racemes, nodes of the raceme bearing several flowers and bracteoles at the base of the calyx 2. Pods are subsessile, tuberculate on the sutures, 3-6 seeded, glabrous when ripe, 2-5 cm long and 1.2-2.2 cm broad.

Pollination: In dolichos bean, self-pollination is predominant. However, cross pollination occurs to an extent of 2-8%. Xylocopa, ants, thrips, butterflies are the main pollinators. Flowers bloom between 11:00 am to 4:00 pm. From the anthers, the release of pollens happen before the anthesis of flowers. Stigma becomes receptive from 8:00 am to 7:00 pm on the day of anthesis.

The stamens are typically organised in a 9+1 (diadelphus) configuration, featuring uniform anthers, a sessile ovary, an incurved style, and a terminal glabrous stigma. Seed colours vary from white, ochreous with black dots, to black with white dots besides uniformly brown or black. Flower blossoms typically 2 days after anther dehiscence between 6:30 a.m. and 7:00 a.m. The redundant floral nectarines located at the corolla's base entices bees, flies and ants. However, intense insect activity may hinder the growth of wing petals, leading to stamen and stigma exertion. Hyacinth beans, primarily self-pollinating but experiencing insect-mediated cross-pollination to a degree of

6-13%. The optimal timing for emasculation of dolichos flowers is between 4:30 p.m. and 5:30 p.m., while hand pollination is most effective between 7:00 a.m. and 9:00 a.m. the following morning. Pollen remains fertile during anthesis day and viable for 42 hrs at room temperature (28.5°C) with 85-90% RH. Extended viability of up to 50-60 hrs is achievable when pollen is stored in a cold environment such as a refrigerator.

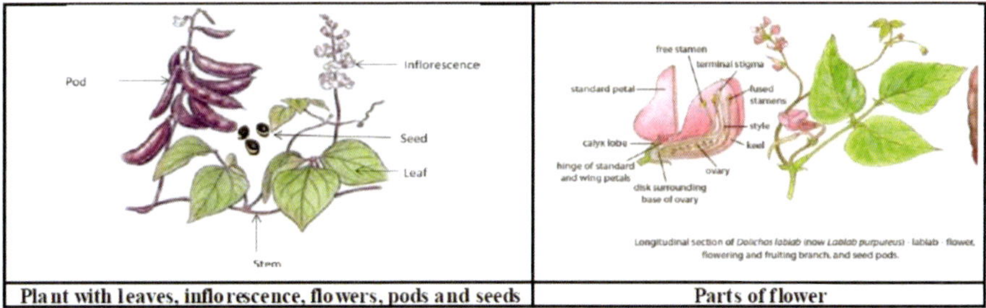

Plant with leaves, inflorescence, flowers, pods and seeds	Parts of flower

Breeding Objectives

- High-yielding, early maturing
- Photo-insensitive
- Nutritious for food and feed, drought-tolerant
- Tolerant to mosaic, anthracnose, rust and pests like pod border and bruchid
- Nutrient rich and/or low anti-nutritional genotypes

The utilization of genomic tools like DNA markers in dolichos bean breeding is in its early stages due to their unavailability. However, sequence-independent marker systems such as RAPD and AFLP have been employed to identify and analyze genetic diversity among germplasm accessions and breeding lines. Unfortunately, the data derived from these markers lack reliability because of their inconsistent reproducibility. Therefore, researchers prefer sequence-dependent markers such as Simple Sequence Repeat (SSR) and Single Nucleotide Polymorphism (SNP) due to their straightforward inheritance patterns, adaptability for automation, and high reproducibility. These markers are advised for their ability to provide more dependable genetic information in dolichos bean breeding programs.

Breeding Centers

- ICAR-IIHR, Bangalore
- UAS, Bangalore
- ICAR-IARI, New Delhi

- Dr. BSKKV, Dapoli
- TNAU, Coimbatore
- CSAUA&T, Kanpur
- MPKV, Akola
- JNKV, Jabalpur.

Varietal achievement: The ICAR-Indian Institute of Horticultural Research (IIHR) in Bengaluru has made significant strides in plant breeding by successfully incorporating traits such as photo-insensitivity and determinate growth from *Lablab purpureus* var. *lignosus* (specifically, Hebbal Avarai 3, a pulse-type Dolichos) into the genetic background of *Lablab purpureus* var. *typicus* (known locally as Kanupu Chikudu, a highly esteemed vegetable type dolichos bean variety). This effort has led to the development of two bush-type vegetable dolichos varieties known as Arka Jay and Arka Vijay, which are well-suited for year-round cultivation. ICAR-IIHR has also developed seven photo-insensitive pole-type varieties of dolichos and five photo-insensitive bush-type varieties specifically tailored for vegetable cultivation purposes.

List of Varieties Available in India Bean

Sl. No.	Name of the Varieties	Salient features	Institute	Images of the Varieties
1.	Arka Adarsh	• Photo-insensitive and Pole type variety. • Pods are medium long, slightly thick, broad and dark green colored. • Pod Yield: 41.0 t/ha in 120 days.	ICAR-IIHR, Bengaluru	
2.	Arka Krishna	• Photo-insensitive and Pole type, early variety. • Pods are borne in clusters and dark green colored. • Pod Yield: 30.0 t/ha in 120 days.	ICAR-IIHR, Bengaluru	
3.	Arka Pradhan	• Pole type and photo-insensitive variety. • Pods are green in colour, smooth and shiny with an undulating surface. • Pod Yield: 35.0 t/ ha in 120 days.	ICAR-IIHR, Bengaluru	
4.	Arka Visthar	• Pole type and photo-insensitive variety. • Pods are thick, long, very broad and dark green coloured. • Pod Yield 37.0 t/ha in 120 days.	ICAR-IIHR, Bengaluru	
5.	Arka Bhavani	• Pole type and photo-insensitive variety. • Pods are wavy, slender and dark green coloured.	ICAR-IIHR, Bengaluru	

6.	Arka Prasidhi	• Pole type and photo-insensitive variety. • Pods are dark green, flat, long and slightly curved. • Resistant to rust.	ICAR-IIHR, Bengaluru	
7.	Arka Swagath	• Pole type and photo-insensitive variety and suitable for year-round cultivation. • Pods are medium long and light green and suitable for Karnataka.	ICAR-IIHR, Bengaluru	
8.	Arka Amogh	• Bush-type photo-insensitive. • Pods are green, wavy, medium long and ready for harvest in 55 days.	ICAR-IIHR, Bengaluru	
9.	Arka Sambhram	• Bush-type photo-insensitive. • Pods are light green, flat, medium long, ready for harvest in 55 days.	ICAR-IIHR, Bengaluru	

10.	Arka Soumya	• Bush type, photo-insensitive. • Pods are medium long, slender, wavy, and ready for harvest in 55 days. • Suitable for growing in Andhra Pradesh	ICAR-IIHR, Bengaluru	
11.	Arka Jay	• Bush type, photo-insensitive. • Pods light green, long and slightly curved, without parchment. • Vegetable type with excellent cooking qualities. • Tolerant to low moisture stress.	ICAR-IIHR, Bengaluru	
12.	Kashi Haritima (VRSEM-8)	• The average yield of green pods during the fruiting period, from the last week of November to the first week of March, is 38 t/ha • This represents a 28.23% increase in yield compared to the national check variety Swarna Utkrisht.	ICAR-IIVR, Varanasi	
13.	Kashi Khushal (VRSEM-3)	• With a cluster-bearing habit, this variety exhibits very early fruiting. • The pods are harvested at 95-100 days after seed sowing. • The average yield of green pods is 35.7 35.7 t/ha.	SVRC in 2018 in UP	

14.	Kashi Sheetal (VRSEM-11)	• The variety achieves an average yield of 35.9 t/ha of green pods, marking a 30.6% increase compared to the national check variety Swarna Utkrisht. • In field conditions, this variety has demonstrated tolerance to DYMV.	SVRC in 2018 in UP	
15.	VRSEM-6	• High yield was reported in centers like Coimbatore, Raipur, Kalyani, Parbhani, Bengaluru, Portblair and Barapani. • The average pod yield is 36.9 t/ha	ICAR-IIVR, Varanasi	
16.	VRSEM-186	• This variety exhibits tolerance to cold temperatures (less than 6°C) and also demonstrates resilience to DYMV in field conditions. • The average yield of green pods stands at 34.1. q/ha, representing a 22% increase compared to the national check variety Swarn Utkrisht.	ICAR-IIVR, Varanasi	
17.	VRSEM-501	• Average yield is 343q/ha green pod	ICAR-IIVR, Varanasi	

18.	VRBSEM-3	• Hybrid developed from VRSEM-11(Pole Type) and Arka Vijay (Bush type) followed by single plant selection. • The average pod yield of 36 t/ha with high protein (93.5mg/g) and carotenoid(0.212mg/g). • This variety exhibits field tolerance to DYMV throughout the cropping season.	ICAR-IIVR, Varanasi	
19.	VRBSEM-9	• Hybrid developed from VRSEM-11(Pole Type) and Gomachi Green (Bush Type) followed by single plant selection. • The variety has an average pod yield of 35 t/ha, featuring notable levels of phenol at 1.81 mg/g and chlorophyll at 1.29 mg/g. • The variety exhibits field tolerance to DYMV throughout the cropping season.	ICAR-IIVR, Varanasi	

20.	VRBUSHSEM-14	• The variety was developed through interspecific hybridization of Kashi Sheetal (pole type) and Konkan Bhushan (bush type), followed by the selection of individual plants. • The plants exhibit a bushy growth pattern, reaching heights of up to 45 cm, and take 51 days to achieve 50% flowering. • The fresh, edible pods contain protein at 77.6 mg/g FW, phenol at 1.36 mg/g FW, and proline at 1.42 ug/g FW. • This line exhibits tolerance to DYMV during the fruiting period.	ICAR-IIVR, Varanasi	
21.	VRBSEM-18	• This variety was developed via interspecific hybridization involving Kashi Sheetal (pole type) and Konkan Bhushan (bush type), followed by the selection of individual plants. • The flowers of this variety are purple, producing light purple twisted pods. • It has a yield potential of q/ha across 4-5 harvests. This line exhibits tolerance to DYMV during the fruiting period.	ICAR-IIVR, Varanasi	

22.	Hebbal Avare-1	• This variety is characterized by its bushy growth and insensitivity to photoperiod. • The pods are small and tender. • It yields t/ha within a span of 90-100 days.	UAS, Bangalore.	
23.	Hebbal Avare-3	• This variety is characterized by its bushy growth and insensitivity to photoperiod. • The flowers are white. • Its pods are green and typically contain 2-3 seeds, with brown, round, and small seeds. • It yields 8-10 t/ha within 90-100 days.	UAS, Bangalore.	
24.	Hebbal Avare-4	• Bush and photo insensitive variety. • The pods are soft and harvested in 5 pickings. • It yields 6 t/ha.	UAS, Bangalore.	
25.	Pusa Early Prolific	• Pole type variety, pods flat, green, narrow, cycle shaped. • Pod length 9.3 cm, width 1.5 cm, weight 3.5 g. • Yield 14 t/ha in 200-215 days.	IARI, New Delhi.	

26.	Pusa Sem 2	• Pole type, pods semi-flat, fleshy, dark green and stringless. • The pods are15-17 cm long. Yields 13-22 t/ha in 200-215 days. • Tolerant to anthracnose, yellow bean mosaic virus, aphids, pod borers and frost.	IARI, New Delhi.	
27.	Pusa Sem 3	• This variety is categorized as pole type, bearing flat, green, fleshy, and stringless pods. • The pods have a length of 15 cm, and the yield ranges from 17 to 27 t/ha within a period of 200-215 days. • It exhibits tolerance to anthracnose, yellow bean mosaic virus, aphids, pod borers, and frost.	IARI, New Delhi.	
28.	Wal Konkan 1	• It is bushy in nature, photo-insensitive and resistant to yellow mosaic virus. • It yields 9-10 t/ha in 110-115 days.	KKVP, Dapoli	
29.	KonkanBhushan (DPLD 1)	• This variety is characterized by its bushy growth, insensitivity to photoperiod, and resistance to yellow mosaic virus. • It yields between 9 to 10 tonnes per hectare within a span of 110-115 days. • Originating from KKVP, Dapoli.	KKVP, Dapoli	

30.	Dasarawal	• It is Pole type and pods dirty green with purple tinge at both boarders. • Pod length is 7.8 cm, width 2.0 cm and weight are 3.2 g. • Yield 7-8 t/ha.	MPKV, Akola	
31.	Deepaliwal	• This variety is classified as pole type, featuring extra-long pods measuring 18.4 cm, which are white in color and exhibit a non-smooth surface due to bulging at each seed. • The pods have a width of 2.7 cm and weigh 1.5 g. It yields between 6 to 8 t/ha within a duration of 200-210 days.	MPKV, Akola	
32.	Phule Gauri	• Stem colour green with white flowers, pod colour whitish green, Pod shape flat and slightly curved, Fruit surface smooth • Tolerant to leaf miner and pod borer, • Moderately resistant to anthracnose and leaf spot under field conditions	MPKV, Akola	
33.	Phule Suruchi	• The pods are straight with a slight curve, displaying an attractive green color with purple tinges along both edges.	MPKV Rahuri	 Dolichos Bean - Phule Suru

34.	Phule Ashwini	• Pods are straight and dark green colour, pold length 10-15 cm, climbing type, stabile for local market	MPKV Rahuri	
35.	JDL 79	• This variety is classified as pole type. Its pods are flat and broad, with a whitish-green color and a parrot green border along the line of seed attachment. • The pods have a length of 11.8 cm, width of 3.6 cm, and weigh 1.5 g. • It yields 5. t/ha over a period of 200 days.	JNKV, Jabalpur.	
36.	JDL 53	• Pole type, pods flat, narrow, small, dull whitish green with purple tinge along the border. • Pod length 7.2 cm, width 1.8 cm, weight 3.75 g. • The pod yield ranges from 10-12 t/ha in 200-220 days.	JNKV, Jabalpur.	
37.	Jawahar Sem 53	• It has been developed through selection. It is extra early and high yielding.	JNKV, Jabalpur.	

38.	Rajani	• This variety is classified as pole type, with narrow oval-shaped pods that have a shining green appearance. • Pod length 10.4 cm, width 1.2 cm, weight 1.78 g. • Yield 7-8 t/ha in 200-210 days.	CSAUA&T	
39.	KDB 403	• This variety is a pole type characterized by its long, narrow, and glossy green pods. • Pod length 12.9 cm, width 1.2 cm, weight 2.0 g. • Yield 5-6 t/ha in 180-210 days.	CSAUA&T	
40.	KDB 405	• This pole type variety features medium-long pods that are narrow, with a dark green band in the middle and light green borders. • The pods measure 9.6 cm in length, 1.3 cm in width, and weigh 1.1 g. • It yields between 3 to 4 t/ha in 180-200 days.	CSAUA&T	
41.	Co 1	• This variety, characterized as pole type, produces green, fleshy pods with a slow development of fibers. • The pods have a weight of 9.7 grams. • It yields between 18 to 20 t/ha in 160-180 days.	TNAU	

Important Companies Involved in Commercial Breeding and Seed Production of Dolichos Bean

- Ashoka Seeds
- Kalash Seeds Pvt. Ltd.

- Ankur Seeds
- Hindustan Seeds Pvt. Ltd.
- Nuziveedu Seeds Ltd.
- Sarpan Seeds
- VNR Seeds
- Empee Agro Tech
- Gloabal Agritech Group of Company
- Vokkal Seeds Pvt. Ltd.

New Developments in Dolichos bean Research and Development:

Ankur Seeds Launches One of the World's First GMS-Based variety of Indian (Dolichos) Bean.

Commercial Practices of Seed Production

- Land designated for seed production should be devoid of volunteer plants.
- Stake the plants after sowing.
- Rouging is done before flowering, during flowering and fruiting to meet field standards.
- Ripe pods are picked and sun-dried.
- The seeds are extracted and dried well, clean seeds are stored in bags in moisture-free condition.

Isolation distance: 50 m for Foundation seed production and 25 m of isolation for Certified seed production is recommended in dolichos bean.

Sowing season and seed rate: January-February or August-September. A seed rate of 35 kg/ha for bush varieties and 4 kg/ha for pole varieties are re-commended.

Seed treatment: Captan/Bavistin @ 2g/kg seeds.

Fertilizers: 25:75:60 kg NPK/ha

Spacing: Bush type: 60 cm ×10 cm

Pole type: 1.0 meter × 0.75 meter

Rogueing: Three rouging are recommended during the seed production of dolichos bean during the following stages:

1. Before flowering
2. At the time of flowering
3. At the time of maturity

Rogues are removed based on leaf color, plant color, pod color and seed color.

Plant protection: Rust, Septoria leaf spot, Cercospora leaf spot, Bacterial blight (caused by *Xanthomonas* sp.), Anthracnose (caused by *Colletotrichum lindemethianum*) and Yellow Mosaic Virus are common. Spraying of Sulfex @ 2g/lt effectively manages rust and Septoria leaf spot while Cercospora leaf spot can be managed by spraying by spraying Dithane M45 @ 2g/litre Mites can be managed by spraying Kelthane 2ml/lt.

Harvesting

Matured and dried pods are harvested followed by the extraction of seeds through threshing.

Seed Standards

Field Standards

Factor	Maximum permitted (%)	
	Foundation	Certified
Off type	0.10	0.20
Designated disease	0.20	0.20

Seed Standards

Factor	Standards for each class	
	Foundation	Certified
Pure seed (min)	98.0%	98.0%
Inert matter (max)	2.0%	2.0%
Other crop seeds (max)	None	None
Total weed seeds (max)	None	None
Other distinguishable varieties (max)	5/kg	10/kg
Germination (min)	75%	75%
Moisture (max)	9.0%	9.0%
For vapour proof container (max)	8.0%	8.0%

Seed processing: The seeds are processed using a round perforated metal sieve grader with perforations measuring 22/64".

Seed Storage

- Dried seeds are stored in cloth bags in cool and dry conditions.
- Best storage temperature:4.5 to 7 ^{0}C, RH 65-70%.
- Longevity: 2-3 years at room temperature under dry conditions.
- Treat the seeds with Thiram @2g/kg and stored in polyethylene bag maintain better germination.

Cultivation Practice for Vegetable Dolichos

Seed Rate

Bush varieties 13 kg/acre

Pole varieties 1.5 kg/acre

Spacing

Bush varieties 60×10 m^2

Pole varieties 1.5×0.75 m^2

Season: The crop can be grown in both *Kharif* and *Rabi* seasons

Bed Preparation: Paired row sowing can be taken up for mid-season varieties by preparing 80 cm beds leaving 40 cm space between two beds.

Fertilizer Dose: 20-30-24 kg NPK/acre along with FYM @10 t / acre during land preparation

Laying of dripline: Install a single in-line drip lateral (16 mm) in the center of the bed, requiring a total length of 3300 m of lateral pipe.

Polyethylene mulching: mulch film of 3300 m length and 1.2 m width and 30-micron thickness is required to cover 80 cm wide bed.

Sowing: Bush and pole varieties: First fortnight of July, October and February. Create narrow, shallow furrows along one side of the rows. Sow seeds in these shallow furrows.

Irrigation: For irrigation, water the plot two days after sowing and subsequently irrigate at intervals of 2-3 days, adjusting based on the soil moisture availability.

Weeding: Maintain weed-free plots with two rounds of hand weeding before earthing up, followed by fortnightly weeding thereafter. Use the remaining half dose of nitrogen during earthing up.

Fertigation: Schedule fertigation with 19-19-19 NPK @ 4g/l for fortnightly intervals for two months.

Foliar Nutrition: Give foliar sprays @ 5 g/l using foliar spray grade fertilizers containing Ca, Mg, Fe, Mn, B, Cu, Zn thrice at vegetative, flowering and pod setting stages.

Staking: Staking is compulsory for pole varieties and it improves crop growth and also reduces disease incidence

Harvesting and Yield: Bush varieties are harvested in 55-60 days and yield 4.0-8.0 t/acre in 90 days. Pole varieties are ready to harvest in 75-80 days and yield 10-16 t/acre in 120 days. Pods can be harvested when they are immature, tender and are grown.

Future Thrust

- Development of photo-insensitive bush and pole-type vegetable dolichos.
- Incorporation of resistance to mosaic, anthracnose and rust in varieties.
- Search for pod borer resistance along with storage problems like bruchids.
- Development of nutrient-rich genotypes/varieties.
- The exploitation of dolichos for multiple uses *viz.,* food, fodder and legume.

Conclusion

The research initiative in legumes in general and Indian bean in particular is very scanty primarily due to its low-value nature. In addition to this, self-pollination makes it less attractive for the development of hybrids due to economic sense. But, of late GMS based hybrids have been developed by few private companies hold significance for future line of work. Anti-nutritional factors in fruits and seeds are one of the limiting factors in expanding these crops in the food basket. Hence, future research strategies should address this issue as one of the emerging areas. With above constraints, Indian Bean is still a potential crop due to its climate resilience adaptation and multiple uses like food, fodder *etc.* in coming years.

References

Nagendra Rai, R K Dubey, Manish Singh and K K Rai (2020). Status, potential and improvement of Indian bean. Indian Horticulture, May-June, 2020. 117-120.

Susmita C, N. Mohan and T. S. Aghora (2020). Breeding for evolution of photo-insensitive pole type vegetable dolichos (*Lablab purpureus* L.) varieties to suit year-round cultivation. Electronic J. Plant Breed. Vol 11(2):633-637.

13

Onion

B.R. Raghu

Division of Vegetable Crops, ICAR-Indian Institute of Horticultural Research Hessarghatta Lake P.O., Bengaluru-560089, Karnataka

Introduction

The onion, or *Allium cepa* var. cepa L., 2n=16, is one of the most important vegetables that are grown and eaten globally (Kiani *et al.*, 2023). It is referred to as the 'queen of the kitchen' because of its wide range of culinary applications (Selvaraj, 1976). It can be eaten raw or prepared in a variety of ways, including as pickles, soups, frying, boiling, and baking (Cattivelli *et al.*, 2021). Apart from being consumed raw, onions also provide excellent raw materials for the processing sector. They can be processed into onion pastes, dehydrated powder, flakes, rings, shreds, and onions in vinegar or brine (Gnanasundari *et al.*, 2023). Additionally, it possesses several health promoting factors such as antihyperlipidemic, anti-diabetic, anti-hypertensive, anti-microbial, immune-protective, anti-obesity properties, and rich in several minerals and vitamins (Pareek *et al.*, 2017). Medicinal value of onion has been known since antiquity; the third or fourth century A.D.'s "Charak Samhita" is one of the medical treatises that mentions it (Ray *et al.*, 1980). Allicin, ajoene, allixin thiosulfinates, sulphites, and other significant biochemical components are found in onion, giving them considerable use and nutraceutical value.

With a production of 23.73 million tonnes from 1.43 million hectares, India is one of the world's largest producers of onion, contributing more than 30% of global production (FAOSTAT, 2020). In India, onions are grown in three distinct seasons: *Kharif*, Late *Kharif*, and *Rabi*. The *Rabi* produces 60% of the total crop, with the remaining 20% each is coming from late *Kharif* and *Kharif* (Lawande and Murkute, 2011; Murkute 2012). In 2020-21, onion was grown on an area of over 16.24 lakh hectares, producing 266.41 lakh tons (NHRDF). The main states that grow onions are Madhya Pradesh, Rajasthan, Andhra Pradesh, Tamil Nadu, Maharashtra, Karnataka, Gujarat, and Bihar. Although grown in varying quantities across the country, Maharashtra is the leader in onion production, accounting for 39.32% of the country's production and 43.27% of its area (NHRDF). Maharashtra is the leading state in terms of output, followed by Madhya Pradesh (45.48 LT), Gujarat (16.57 LT), Rajasthan (13.86 LT), Bihar (13.28 LT), West Bengal (7.47 LT), and Andhra Pradesh (6.36 LT) (NHRDF 2021). With the exception of higher altitudes in the Himalaya and other mountainous regions of the country, the majority of onions grown in India (more than 90% of the total farmed area) are short day onions, which require 10 to 12 hours of daylight for bulb formation.

Despite being the world leader in both total cultivated area of onion and its production, India's productivity of 15.9 t/ha is less than the global average of 20.4 t/ha. India's low productivity is due to several constraints, one of which being the lack of high-quality seeds for planting each year during planting seasons. About 12169.6 tonnes of high-quality seeds are needed each year (at a seed rate of 10 kg per hectare) to meet the planting material requirements of onion-growing farmers. Only 9.6% of the overall annual requirements are met by NHRDF, NSC, ICAR organizations (IARI, IIHR, DOGR), SAUs and other national institutions. Private sectors, unorganized programs, and farmers' own saved seeds meet the remaining demand for seed-planting materials requirements. Farmers' own seeds account for almost 60% of the need for onion seed supplies in the country. These farmers' seeds are mostly genetically uneven, diverse in composition, and have low germination and physical purity, which results in low yield and subpar, unmarketable bulb quality. To attain high onion productivity in India, it is crucial to increase the supply of high-quality seeds, particularly those of improved cultivars and hybrids, by effectively using seed production technologies.

Floral Biology

The initiation of flowering in onion is influenced by several environmental factors. Cool temperature is the main factor responsible for flower initiation, and day-length has no impact on it as it does for bulb development. Bulbs or growing plants with four or more leaves generally start flowering when temperatures are 40°F or below for one week. However, the temperature before and after the 40°F week can affect the induction of flowers. Small seedlings are generally unresponsive to cool temperatures, but the larger the plant, the easier it is to induce flowering. Induction of flowering causes the shoot apex to stop producing leaf primordia and initiate inflorescence. The inflorescence of onions known as an umbel (Figure 1) is an aggregate of many small inflorescences (cymes) of 5 to 10 flowers, and it is borne on a scape. The inflorescence may consist of a few to more than 2000 flowers per umbel (Pike, 1986). Each flower opens in a definite order causing the flowering to be irregular and to last for two or more weeks. The flower stalk, known as the scape or seedstem, which bears the umbel consisting of the spathe and flowers, is a one-internode extension of the stem. The stalk is initially a solid structure, but as it grows, it becomes hollow. The number of flower stalks produced per plant depends on the number of lateral buds that are contained on the stem, which is the compact base plate on the bottom of the bulb. Bulb-grown plants usually produce 6 or more flower stalks, while seed-grown plants usually produce only one stalk.

Each flower is composed of 6 stamens, 3 carpels that are joined together into a single pistil, and 6 perianth segments. There are 3 locules in the pistil, and each of them contains 2 ovules. Furthermore, nectarines can be found in the flower and produce nectar to attract insects for cross-pollination. The flowers are strongly protandrous; the self-pollination is largely absent since anthers shed pollens even before the stigma becomes receptive (Delaplane and Mayer, 2000), and it continues over a period of 3-4 days prior to the time when full length of style is attained, and thus, it depends on insects for cross pollination. Anthesis (flower opening) usually occurs in the early morning (6-7 AM), but it can last until 9-10 AM depending on the weather conditions. Anther dehiscence (release of matured pollen from anthers) is happening between 7.00AM and 5.00 PM and continues the next day, with a peak between 9.30AM and 5.00 PM. Pollen fertility is maximum on the day of anthesis. Thus, stigma becomes receptive 3-4 days after shedding of pollen grains and protandry leading to favor cross pollination. After pollination, the seeds develop. As they mature, the capsules dry and split open at the apex and run down the center of each locule, which allows the seeds to fall free upon maturation.

Fig. 1. Onion inflorescence before **(A)** and after anthesis **(B)**

Pollination Control in Onion

Controlled pollination in highly cross-pollinated crops, like onions, is very challenging since each umbel has several hundreds of tiny individual perfect flowers. Furthermore, the technique of controlled pollination in onions is becomes extremely complex due to presence of male sterility, the sole mode of selfing by geitonogamy, and the prolonged time of anthesis spanning 2-3 weeks in a particular umbel. However, it is crucial to attain controlled mating in onions to understand the genetics of various economically important traits and to achieve desired mating between chosen elite individuals.

Selfing Technique

After the inflorescence appears, place two to three umbels of a superior bulb in a butter paper bag and shake it every day between eight and ten in the morning for a week. This will allow for self-pollination, which may result in superior seed set. The technique is more suitable for development of inbred lines in onion for hybrid development (Figure 2). However, selfing can done on a limited basis because inbreeding depression begins showing in second generation (S2) onwards. In order to overcome the undesirable effects of inbreeding depression, half or full sib matting is followed in advanced generations of inbred development.

Sib-mating

Since onion suffer a greater degree of inbreeding depression after two to three generations of selfing, sib-mating is employed to achieve seed setting in advanced generations. Two-three superior bulbs of closely related perigee (full or half sib) were selected and closely planted. All the umbels were enclosed in insect-proof cages to achieve cross pollination among them in isolation. Due to the lack of insect activity inside the cage, hand pollination should be carried out daily from 8-10 AM during peak flowering to increase percent seed set. During hand pollination, the pollens will be distributed over the umbel periodically using a cotton ball rolled in muslin cloth.

Figure 2. Selfing in onion by butter paper method **(A)** and insect proof cage method **(B)**.

Hybridization Technique

The breeders have two choices to make initial crosses between two selected lines through hybridization. One is hand emasculate the stamens from one line followed by hand pollination with pollen of the selected elite line. Though it is extremely difficult because flowers open on an umbel for 2 or more weeks adds to the problem of making crosses. However, in order to achieve efficient hybridization through hand emasculation, we choose unopened buds that are ready to open the next morning. The remaining flowers in the influence are removed. Selected unopened buds are opened slowly using forceps and anthers are removed. After hand emasculation, the umbels are enclosed in butter paper. After 2-3 days of emasculation, pollens from the selected male plant were collected with cotton ball rolled in muslin cloth and periodically distributed the pollinated over emasculated flowers. The second option is to make what is called a fertile x fertile cross, where two selections are caged together and then pollinated by hand or with insect pollination, like with honey bees. In this method, difficulty of hand emasculation does not exist, and it is very effective only when two selections are different enough to distinguish F_1 hybrid from selfed seeds in the bulb stage.

Massing Technique

Cross pollination in isolation among selected elite lines is achieved by massing technique. 50-60 selected bulbs are close planted and enclosed in insect proof cage to achieve cross pollination among them in isolation (Figure 3). Due to lack of insect activity inside the cage, human hand pollination should be carried out daily in morning 8-10 am for 8-10 days during peak flowering increase percent seed set by periodical distribution of pollen over the umbel with cotton ball rolled in muslin cloth. Otherwise, flies are again the best choice to pollinate the big lots. For that cages may be used up to 100ft long by 12ft wide and bees are used to pollinate among selected lines in isolation.

Figure 3. Massing technique in onion by insect proof cage method.

Breeding Objectives

General Breeding Objectives

Efforts of onion breeders should prioritize traits that contribute to the production of high-quality seed and bulbs. Bulb traits are size, shape, colour of skin and inner scales, single core, skin retention, firmness, dormancy, pungency, and concentration of soluble solids (TSS). In addition, production of commercially acceptable bulbs is dependent on resistance to diseases, pests, and non-bolting (no flowering during bulb production). Vigorous bulbs, synchronous flowering, straight seed stalks, pollinating insects, and ideal drying and cleaning conditions are all necessary for the production of high-quality seed. Thus, breeder should formulate the objectives in the beginning of the programme to achieve desired goal. Some of the breeding objectives in onion breeding programs are listed below.

1. High bulb yield.
2. Superior bulb quality traits (size, shape, colour, firmness, pungency, single core, low splits and doubles and dormancy).
3. Resistance to diseases (mainly, purple blotch, *Stemphylium* blight, basal rot, anthracnose-twister disease).
4. Resistant pests, mainly trips.
5. Resistant to abiotic stresses (drought tolerance, high moisture stress, high temperature stress, salinity, and alkalinity).
6. Development of high yielding varieties capable of producing good quality seeds.
7. Development of varieties suitable for export market.
8. Development of F_1 hybrids based on stable CMS system in short day onion background.
9. Short duration in *Kharif* season onion.
10. Longer shelf life in *Rabi* season onion.
11. High total soluble solids (TSS) important in white onion for dehydration industries, besides skin retention and high dry matter recovery.

Current Industrial Preference

However, the industrial priorities are different from general breeding objectives that are followed in institutional breeding programmes. Some of the current industrial priorities in onion are listed below.

1. **Common red onion**: Medium bulb size (4.5 to 6.5cm diameter, 100-120 gm bulb weight), thin and attractive outer skin, bright and shining outer skin with medium to deer red colour. High percent of single core bulbs, low percentage doubles and split bulbs. Bulbs with longer shelf life suitable for storage up to 6 months. Stable CMS sources under different elite genetic background is current more demand by the onion seed industries for production of different combination hybrids.
2. **White onion:** Mainly white onions are grown for requirement of dehydration industries. Bigger bulb size of >6.0 cm diameter is preferred which are suitable for dehydration process. High TSS of more than 18% are preferred by the dehydration industries. Longer skin retention and high dry matter recovery is another trait required by dehydration industries in white onions.
3. **Multiplier onion:** True seed multiplier onion varieties with short duration of less than 90 days to harvest. More than 5-6 bulblets per plant preferred for gaining higher yield.

4. **Rose onion:** Small size bulbs (3.5 to 4.5 cm diameter, 30-35 gm weight bulb) which are suitable for export. High percentage of uniform bulbs with less splits and doubles bulbs. Attractive skin with deep pink to deep purple colour bulbs.
5. **Yellow onion:** Mostly preferred for export purposes. More uniform and big size bulbs with high TSS and low pungency are preferred by the export markers in yellow onions.

Varietal Achievements

In India, the organized research on onion began in 1960 at Nashik and later at IARI, New Delhi, IIHR, Bengaluru, and NHRDF, Nashik (Lawande *et al.* 2009). NHRDF played a vital role in establishing systematic research on onion, resulting in the development of several high yielding onion varieties and production methods, ultimately leading to a rise in onion production in India. In the early nineties, onion research in the country was strengthen by State Agricultural Universities (SAUs) and All-India Coordinated Research Project (AICRP). They developed new onion varieties for different agro-climatic zones and improved agronomical practices. In 1994, ICAR established the National Research Centre on Onion and Garlic (NRCOG) in Nashik with the goal of increasing productivity with a particular focus on onion research. In 1998, the centre was moved to Rajgurunagar, and in 2009, it was upgraded to become the Directorate of Onion and Garlic Research. Currently research on onion is being aided by several ICAR institutes and a few SAUs, such as CCSHAU Hisar, PAU Ludhiana, and MPKV Rahuri. As a result, over 60 varieties of onion including 2 F1 hybrids and 6 varieties of multiplier onion have been developed and released in India so far. Although onion is mostly grown during the *Rabi* season, it is also grown on a sizable area during the *Kharif* and Late *Kharif* seasons. Thus, cultivars are bred specifically for various seasons and geographical areas within the nation. Baswant-780, N-53, Agrifound Dark Red, Arka Kalyan, and Bhima Super are popular improved varieties for the *Kharif* season; Baswant-780, Bhima Red, Bhima Shakti, Phule Samarth, and Agrifound Light Red popular for the late *Kharif* season; and N-2-4-1, Arka Niketan, Bhima Raj, Bhima Red, Bhima Kiran, Bhima Shakti, Agrifound Light Red, Pusa Red, Pusa Madhawi, etc. most preferred varieties for the rabi season. There are white coloured varieties which can be grown during rabi season are Phule Safed, Pusa White Round, Pusa White Flat, Bhima Sweta, Agrifound White, Punjab Selection, Udaipur – 102 and Bhima Shubra for kharif season. List of popular released onion varieties of onion for different season and regions of country are given as follows (Table 1):

Table 1. Ruling onion varieties and hybrids with respect to different segments

Variety/Hybrid	Organization	Salient features
		Kharif season
Arka Kalyan	ICAR-IIHR, Bengaluru	Selected from local collection of Kalwantaluka of Maharashtra, Bulbs shape – globose; Dark red colour; Average yield: 33t/ha; Suitable for *kharif* season.
Arka Pragati	ICAR-IIHR, Bengaluru	Improvement over local collection IIHR-149 of Nashik; Colour of bulb attractive pink, Globular shape, Uniform size with thin neck, highly pungent; Early in maturity 95 to 100 DAT; Average yield is 20 t/ha. Suitable for *kharif* season.
Arka Lalima	ICAR-IIHR, Bengaluru	High yielding F_1 hybrid developed through heterosis breeding (CMS system). Medium to big sized bulbs with globe shape and firm texture. Field tolerance to diseases and pests. Suitable for *kharif* and *rabi*. Bulb yield: 47 tonnes /ha.
Arka Kirthiman	ICAR-IIHR, Bengaluru	High yielding F_1 hybrid developed through heterosis breeding (CMS system). Medium to big sized bulbs with globe shape. Developed through heterosis breeding (MS based). Field tolerance to diseases and pests. It is suitable for *kharif* and *rabi* and has yield potential of 45 tonnes /ha.
Arka Bheem (Syn 6)	ICAR-IIHR, Bengaluru	It is a tri-parental synthetic variety with Red to pinkish red elongated globe shaped bulbs. Average bulb weight is 120 g. It yields 47 tonnes /ha in 130 days.
Pusa Red	IARI, New Delhi	Average bulb weight: 70 – 90 g; Bronze red in colour; Flat to globular in shape; Good in storage. Maturity: Average yield: 25 t/ha
Bhima Super	ICAR- DOGR, Rajgurunagar	Suitable for *kharif* and late *kharif* cultivation in Maharashtra, Karnataka, and Gujarat; Medium red colour; Round with tapering neck; Average yield: 26-28 t/ha in *kharif* and 40 to 45 t/ha in late *kharif.*
Phule Samarth	MPKV, Rahuri	Developed through selection from Sangamner; Bulbs are Dark red in colour, Yield potential of 30 – 31 t/ha in *kharif* and 50– 52 t/ha in late *kharif.*
Agrifound Dark Red	NHRDF, Nashik	Selected from local collection of *kharif* onion grown at Nashik; Yield: 30-40 t/ha; Average in storage; Recommended for *kharif* season all over the country
Baswant - 780	MPKV, Rahuri	Bulbs attractive red in colour; Average yield: 20-25 t/ha; Suitable for *kharif* season in Maharashtra.
N-53	Agriculture Dept., Maharashtra	Suitable for *kharif* season all over country; Bulbs globular in shape; Scarlet red in colour; Average yield is 15-20 t/ha.
Bhima Dark Red	ICAR- DOGR, Rajgurunagar	Average marketable yield: 22-24t/ha; Matures in 100-110 days, Dark red in colour flat glob shape bulb.

Bhima Red	ICAR- DOGR, Rajgurunagar	Bulbs attractive red colour; round shape. Maturity -115-120 DAT for late *kharif* and *rabi*.
Bhima Raj	ICAR- DOGR, Rajgurunagar	Bulbs dark red in colour; oval shaped; TSS: 10 to 110 Brix.; Suitable for *kharif* and late *kharif* Average yield -25-30 t/ha; yield potential: 40-45 t/ha during late *kharif*.
	***Rabi* Season**	
Arka Niketan	ICAR-IIHR, Bengaluru	Bulb shape - globular with thin neck; Attractive light red colour, Average yield is 34 t/ha; Recommended for *rabi* season.
Bhima Shakti	ICAR- DOGR, Rajgurunagar	Average marketable yield- 22-24t/ha; maturity- 100-110DAT; Dark red in colour; flat globe shape bulb.
Pusa Madhavi	IARI, New Delhi	Selected form local collection of Muzaffarnagar, Light red in colour; Flatish round in shape; Good Average yield: 30 t/ha; Recommended for *rabi* season.
Bhima Kiran	ICAR- DOGR, Rajgurunagar	Average marketable yield: 28-32t/ha
Bhima Light Red	ICAR- DOGR, Rajgurunagar	Recommended for Maharashtra and Tamil Nadu for rabi season. Average marketable yield is 38.5 t/ha; matures in 115 days; TSS: 130 Brix. Almost free from doubles and bolters.
Agrifound Light Red	NHRDF, Nashik	Bulbs globular in shape; light red colour; Average yield: 30-32 t/ha
	White Onion	
Arka Yojith	ICAR-IIHR, Bengaluru	It is white onion variety developed for dehydration. It is white in colour, flat globe shape, bulb weight 60-80g, TSS 18-20%, dry matter content 18-20%, bulb weight 60-80g, diameter 4.5-5cm with bulb yield of 25-30 tonnes /ha in 110-120days.
Arka Swadista	ICAR-IIHR, Bengaluru	White onion developed for fermented preservation with TSS 20^{0}brix.Developed through pedigree breeding. Bulb yield is 30 tonnes /ha.
Bhima Shweta	ICAR- DOGR, Rajgurunagar	Average marketable yield: 18- 20t/ha (*kharif*), 26-30t/ha (*rabi*); matures in 110-120 days, White in colour, suitable for processing.
Bhima Shubhra	ICAR- DOGR, Rajgurunagar	Average marketable yield: 18-20t/ha (*kharif*), 36-42t/ha (late *kharif*); White in colour; suitable for processing.
Bhima Safed	ICAR- DOGR, Rajgurunagar	Average marketable yield: 18-20 t/ha; Suitable for processing.

Agrifound White	NHRDF, Nashik	Selected from local germplasm of white onion grown in Nimad area of MP; Bulbs - globular in shape with tight skin; Silvery attractive white colour; Averages yield: 20-25 t/ha; Good for dehydration suitable for *rabi* season.
Udaipur-102	RAU, Udaipur	Bulbs white in colour; Round to flat in shape; Maturity: 120 DAT Average yield: 30-35 t/ha.
	Yellow Onion	
Arka Pitamber	ICAR-IIHR, Bengaluru	Notified during 2006, Released for cultivation in Karnataka; Bulbs yellow in colour; globe shape; very firm, thin neck, mild pungent; good keeping quality; It is suitable for export to European countries, USA, and Japan; Yield potential: 35 - 37 t/ha.
Arka Sona	ICAR-IIHR, Bengaluru	Arka Sona is a yellow onion developed for export market for Asian countries. Developed by pedigree method. The bulb yield is 45 tonnes /ha in 120 days.
Phule Suwarna	MPKV, Rahuri	Bulbs yellow in colour; Suitable for export Less pungent; Suitable for late kharif and *rabi* season. Average yield: 24 t/ha.
	Rose onion	
Arka Bindu	ICAR-IIHR, Bengaluru	Rose onion variety, developed from local collection for Chickballapur area of Karnataka particularly for export; Bulbs deep pink in colour; Bulbs flattish globe in shape; Free from bolters and doubles; TSS: 14-16^0 Brix; Maturity: 100 DAT; Average yield: 25t/ha
Arka Vishwas	ICAR-IIHR, Bengaluru	Rose onion developed for export market (Asia) developed through selection. Bulb yield is around 30 tonnes /ha.
Agrifound Rose	NHRDF, Nashik	Pickling type variety Exclusively for export; Bulbs flattish round; Deep scarlet red in colour; Average yield: 19–20 t/ha;
	Multiplier Onion	
Arka Ujjwal	ICAR-IIHR, Bengaluru	True seed multiplier onion developed for export market developed through pedigree breeding with TSS 180brix.Bulb yield potential is 30 tonnes /ha.
Agrifound Red	NHRDF, Nashik	Number of bulblets /cluster 5.79 (average); weight of single bulblet - 8.85 g; Average yield- 18-20 t/ha.
Co-1	TNAU, Coimbatore	Medium sized bulblets of red colour; 7-8 bulblets / plant; Average weight of bulblets is 55-60 g/ clump; Average yield 10 t/ha.
CO-3	TNAU, Coimbatore	Bulblets pink in colour and 8-10 bulblets per plant; weighing 75 g per clump; Average yield -16 t/ha.

Figure 4. Popular ruling varieties of common red onion recommended for *Kharif* season (***Source:*** www.dogr.gov.in; www.icar.org.in)

Figure 5. Popular ruling varieties of common red onion recommended for *Rabi* season (*Source*: www.dogr.gov.in)

Figure 6. Hybrids and synthetics in common red onion

General Practice of Seed Production

Onion seed production is done in specialized localities. It involves two seasons of activity. Mother bulbs will be produced in the first season. In the following season, seed production will be taken up from the saved mother bulbs. Thus, climate and soil should be suitable for both bulb and seed production. Sometimes bulb production can be done in one place and seed production can be done in another place where favourable conditions are available. In southern parts of India, such as Karnataka, onion seed production is a one-year activity, where bulb production is done in the *Kharif* season and seed production is carried out in the following *Rabi* season. In northern India, onion seed production is carried out over a two-year period, with bulb production taking place during the first year's *Rabi* season and seed production during the subsequent *Rabi* season.

Climate

Onion crops cannot thrive under extreme climatic conditions. The crop suffers substantial damage due to heavy rainfall, high temperatures, and extreme cold. Clear climate is ideal for the crop. Although slightly lower temperatures and shorter day length are sufficient in the initial stage of the crop, slightly higher temperatures and longer day length are required in the bulb development stage. After bulb planting, for proper vegetative growth, a mildly cold temperature of 20°C with lower relative humidity is required, followed by an increase in temperature of 25-30°C in the later stages of the crop. Long rainy periods, heavy dews, and fogs favours incidence of purple blotch, *Stemphylium* blight and downy mildew. During flowering period, harvesting, curing, and threshing of seeds, clear bright sunny days are required which may also favourable for insect activity.

Planting Season

Usually under Northern and Eastern part of India mother bulbs are produced between October to January and these bulbs are stored from February to September. The mother bulbs are planted during October and seeds are produced at the end of January or February. Under southern part of India (Bangalore) where climate is mild with the average temperature of 27° C and moderate rains during *Kharif* season. The mother bulbs are produced during June–September months and bulbs are stored during October month. The mother bulbs are planted during November month and seed crop is harvested in the month of March. There is no dormancy in the bulbs. Proper drying and curing of bulbs will reduce the problems associated with the bulb dormancy.

Soil

Onion can be grown in a variety of soil types, including sandy loam, clay loam, silt loam, and heavy soil. Light sandy loam soils are ideal for mother bulb production. Heavy, cooler soils are suitable for producing high-quality seeds because they can hold more water and minimize the issue of lodging because the soil holds plants firmly.

Bulb Crop (First season-*Kharif* or first year-*Rabi*)

Nursery Management

It is the most important practice to produce healthy seedlings suitable for transplanting. The nursery field should be ploughed deep to break clods and bring to fine tilth. Area of 0.05 hectare is enough for getting seedlings to transplant in one hectare area. At the time of the final ploughing, 500 kg of well-decomposed farm yard manure (FYM) and 1.25 kg of *Trichoderma herzianum* should be applied, and the soil should be thoroughly mixed. Prepare raised seed beds of 25ft long, 4ft wide and 10-15cm height to raise healthy seedlings (Figure 7). With a good quality seeds and proper nursery management practices, 5-7 kgs of seeds are sufficient to raise the seedling for one hectare area. Seeds should be treated before sowing with thiram/ captan/ carbendazim @ 2-3 g/kg of seed to avoid several soil or seed borne diseases. Sow the seeds in lines at 50-75 mm apart at a depth of approximately 1 cm to ensure proper growth, easy removal of seedlings for transplanting, proper weeding, spraying etc. For uniform germination, cover the seeds after sowing with fine soil or vermicompost or powdered farmyard manure, then lightly water. Apply NPK fertilizer at 2:1:1 kg/500 m^2 before sowing and 1 kg N 20 days after sowing. Light irrigation should be given in the morning and evening for initial days to keep top layer soil moist until get good germination. Later irrigation should be applied according to the need. Seedlings will be ready for transplanting in about 6-8 weeks.

Figure 7. Raised bed nursery management- raised bed preparation **(A)**, line sowing in raised nursery beds **(B)**, water management in nursery **(C)** and healthy seedling **(D)**.

Land Preparation for Main Field

Main field is prepared by ploughing, harrowing, and disking to a fine tilth suitable for seedlings transplanting in bulb crop and bulb planting in seed crop. The organic manure of 25 t/ha is incorporated into the soil. Planking should be done for proper levelling. The field is divided into beds and channels. Flat beds should be prepared to be 1.5-2.0 m wide and 4-6 m long. *Kharif* crop should always be grown on raised beds and well-drained soils (Figure 8).

Fig. 8. Raised bed preparation in the main field for onion seedling transplanting.

Transplanting

Age of seedling at the time of transplanting is most important for better crop establishment in the main field. In general, the kharif season nursery should be transplanted 35-40 days after sowing. However, for late kharif, the recommended duration is 40-45 days after sowing, and for rabi it is 50-55 days. Before transplanting, it is recommended to soak the seedlings in a solution of carbendazim (1gm/litre water) and carbosulphan (2ml/litre water) for two hours, ensuring only the root portion is completely immersed (Figure 9). This helps to reduce early incidence of pests and diseases.

Fig. 9. Seedling treatment before transplanting **(A)** and transplanting in the main field **(B)**.

Spacing

The optimal spacing for planting is 15 cm between rows and 10 cm between the plants (Figure 9). As each seedling produces a single bulb, maintaining a sufficient number of plants per unit area is crucial to achieve maximum yield. Additionally, a low population density can lead to an increased number of double bulbs, thick-necked bulbs, and non-uniform bulbs.

Manures and Fertilizers

Approximately 15 tonnes of FYM or 7.5 tonnes of poultry manure or vermicompost per hectare should be incorporated at the time of final land preparation. The recommended doses of chemical fertilizers for *Kharif* are 75:40:40:30 kg NPKS and for late *Kharif* and *Rabi* is 110:40:60:30 kg NPKS. At the time of planting, one third of the recommended dose of nitrogen (N) and a full dose of phosphorus pentoxide (P_2O_5), potassium oxide (K_2O), and sulphur (S) are applied. The remaining two thirds of nitrogen is then split into two equal halves and applied at 30 and 45 days after planting. This method ensures optimal nutrient distribution and absorption. Besides, application of water-soluble micronutrients as areal spray or drenching at rootzone is beneficial based on soil test reports.

Weed Management

Onion crop competes poorly with weed population. Thus, management of weeds is essential for successful cultivation of onion. To increase yields of marketable bulbs, weed growth must be controlled from the beginning. Due to scarcity and high cost of labours, it is inevitable to rely on chemical weed control methods along with some cultural practices. Effective weed control can be achieved by applying Oxyflurofen at 23.5% EC (1.5 -2.0 ml/L) or Pendimethalin at 30% EC (3.5-4 ml/L) either before or at the time of transplanting, and then perform a single hand weeding 40–60 days later.

Irrigation Management

Since onion is shallow rooted bulb crop, they respond very well to better water management. The crop should be regularly irrigated throughout the growing season to realize its maximum yield potential. Generally, onion crop needs to be irrigated during transplanting, three days following, and then at 8–12-day intervals, depending on the season, soil moisture status, climatic conditions, and soil type. Irrigation should be stopped when the crop attains maturity (10-15 days before harvest) and about 5% top starts falling which helps in reducing the rotting during storage.

Harvesting

Crop should be harvested when more than 50-70 per cent of the tops are fall down (neck fall). Harvest the crop by hand-pulling of bulbs along with leaves from the beds followed by field curing for about 3-5 days by ensuring not exposing the bulbs directly to the sun (Figure 10). After curing, leaves should be clipped leaving a 2.0-2.5 cm stalk above the bulbs. The roots of the bulbs should be left intact after harvest. Small size bulbs, twin bulbs, long necked bulbs, bolter bulbs and damaged bulbs if any should be discarded. The medium size bulbs weighing 50-80 g are selected and stored for next seed crop. The selected good quality bulbs are stored at the temperature of 11-12° C and the relative humidity 70-75 %. This helps in uniform early flowering with the heavier seed yield. The bulb storage at high temperature may delay or inhibit flowering.

Figure 10. Onion crop harvesting **(A)**, field curing **(B)**, bulb sorting **(C)** and storage **(D)**.

Seed Crop (Second season-*Rabi* or Second year-*Rabi*)

Raising the seed crop is similar to bulb crop such as land preparation, weed management, irrigation management, pests, and disease management and etc., However, seed crop has few specific requirements that are discussed in the following section.

Selection of Mother Bulbs

Bulbs produced in an area of one hectare is sufficient to plant 3-5 ha for the seed production. The bulbs selected for seed production are usually referred to as mother bulbs. The weight of the bulb has markedly influenced seed production in onion. Increases in bulb weight have increased the yield of seeds. Although an increase in weight and size of bulbs results in higher seed yield, very large size bulbs (> 90 g) require a very high seed rate (60 q/ha), which is not economical. Large size bulbs (3-4 cm diameter) and weighing >90 grams may yield 10.00 q/ha. The medium size, uniform and true to type of desired variety bulbs must be selected for seed crop. Bulbs are roughed during this stage according to their varietal traits for bulb colour, shape, and size. Bulbs will be screened for internal qualities and diseases before planting. Medium sized bulbs require about 15 q of bulbs for planting one hectare area of seed crop. The ideal bulb selection should be based on following criteria;

1. Bulb should be true to type
2. Uniform colour
3. Uniform shape
4. Medium to large in size (45-60mm diameter), preferably more than 80g of bulb weight
5. Preferably single centre

Planting of Mother Bulbs

The top 1/3 of the bulb is cut and smeared with 5% Blitox paste and dried in shade (Figure 10). These bulbs are planted halfway on the ridge at a distance of 30 cm in the field (spacing 45 × 30 cm) (Figure 10). The bulbs are planted in such a way that the cut portion of the bulb is partially visible. Deep-planted bulbs may rot because of soil crust formation on the bulbs, which prevents the sprouting of bulbs. The soil moisture should be maintained at the optimum level until all the bulbs in the field sprout properly. Close spacing of bulbs results in high seed yields, but it also favours diseases. The ideal planting would be 45 cm between the rows and 30 cm between the plants. Bulbs are planted normally during October-November for seed production.

Figure 11. Mother bulbs treated with Blitox on 1/3rd cut portion **(A)**, Planting of mother bulbs for raising of seed crop **(B)**.

Nutrients Management

It is essential to have an optimal level of nitrogen in the initial stages of bulb planting as it is required for proper vegetative growth, and it will continue to be required in more quantity until the differentiation of seed stock. In the later stages phosphorous and potassium requirement is more which facilitates flowering and seed set. The recommended dose of fertilizer for seed crop is 100:50:50:50 NPKS/ha. One-third of the recommended nitrogen (N) and the full quantity of phosphorous (P), potassium (K), and sulphur (S) should be applied as a basal dose before planting. The remaining quantity of nitrogen should be applied in two equal parts, at 30 and 45 days after planting. During flowering, it is not advised to apply N, K, and S because they make the flower nectar unattractive to bees. The seed yield and quality was found to enhance with the application of micro nutrients like Molybdenum., Boron, Zinc and Manganese. It is beneficial to use soil testing-based approaches when applying micronutrients. Commonly, a micronutrient mixture of 5 ml/l is recommended at 45, 60, and 75 days after transplanting. Boron can be applied after flowers open to improve pollen germination (Mahajan *et al.* 2017).

Isolation and Rouging

Onions are a highly cross-pollinated crop, so more care is needed to isolate them from other varieties of onion seed crops and bolters of bulb crops. Isolation distance of two km between cultivars and three km between the cultivars of different bulb colour is require to be maintained. Generally, pure seed crops require four rouging before flowering, during seed stock development, during flowering, and at harvesting. The entire rogue plant should be pulled up, and the flower heads should be cut off and destroyed by burying them. Because, the flower heads which are dropped on the ground or at the end of the row may continue to flower and shed pollen for several weeks.

Supplementary Pollination

Although, onion flowers are hermaphrodite, the self-pollination seldom occurs due to protandrous behaviour. Insects are the major pollinators in the onion. Nectars secreted from onion flowers attacks variety of insects and facilitates cross pollination in onion. Thus, four to six colonies of honeybees should be placed in one hector seed plot at the time of flowering to facilitate good pollination and produce quality seeds (Figure 12). Application of any insecticides should be avoided after flowering to ensure good honey bee population in the seed crop plot.

Figure 12. Onion seed crop with honeybee colony box.

Pests and Disease Control

Onion seed crop is mainly affected by major diseases such as Purple Blotch, *Stemphyllium* Blight and Onion Yellow Dwarf. Selection of healthy bulbs and rotation of onion crop with non-related crops once in 2-3 years will helps in effectively managing these diseases. application of fungicides such as Mancozeb (0.25%), Hexaconazole (0.1%) and Tricyclazole (0.2%) can effectively control purple blotch. Alternative spray of these fungicides may be taken up at the intervals of 15 days, if disease continued to persists in the field. However, application of these fungicides should be stopped after flowering. The safe control measures such as biological control agents namely *Fluorescence pseudomonas*, *Bacillus subtilis*, *Trichoderma viridae*, *T. harzianum* can be used after flowering. Yellow dwarf virus is found to be transmitted by mechanical means as well as by insect vectors. Thus, selection of healthy bulbs in the beginning of planting, and uproot and burning of disease plants immediately after disease appearance would effectively manage yellow dwarf virus. Trips and mites are the common pests of onion seed crop. Trips can cause improper development of inflorescence and induces sterility and poor seed set. It is reported that about 50-60% loss can be incurred due to trips attack in onion seed crop (Mahajan *et al.*, 2017). Thus, before initiation of flowers application of insecticides such as Fipronil, Spinosad, Profenophos and Carbosulphan can manage early incidence of these pests. The sticker like triton/ sandovit should also be mixed in the spray solution to ensure good spray distribution on the foliage. Generally, application of any agrochemicals should be avoided during the morning hours to ensure good honeybee activity around seed crop plot.

Harvesting, Drying, and Threshing

Seed crop gets ready for harvest in about 110-120 days. However, it do not mature synchronously. When seeds are matured and ready for harvest, the seed stock turn from green to yellow color. Umbels also turn yellow with opening of few fruits (capsule) at the centre of umbel. Umbels should be harvested immediately in staggered manner as and when umbels mature in the plants. Around 3-4 harvestings are required to complete full harvest of the seed crop. Generally, harvesting is done manually by cutting the umbels by sickles. While harvesting, umbels should be supported firmly by hands to avoid shuttering loss of seeds. After harvest, umbels are dried in thresh yard by ensuring free from any contamination and mechanical mixtures (Figure 13). Umbels should be spread in shallow layers of not more than 20 – 30 cm height and are exposed to the sun for 4-5 days. Regular turning of heaps is followed to ensure uniform drying of umbels. After complete drying, umbels are beaten gently with wooden mallets or sticks to remove the husks. Seeds are cleaned by winnowing and graded using sieves. Only high-quality seed are packed after cleaning and removing any weed seeds, light seeds, and chaff present in the seed lot.

Packaging and Seed Storage

Bring down the seed moisture to 8% prior to bagging and package. Usually, onion seeds are packed in cloth or jute bags. However, onion seeds loos viability very quickly within 12 months of storage, if they are packed improperly. To extend the viability, it is recommended to pack in a polythene bag (400 gauge) of 1 to 5 kg capacity. But ensure that the seed moisture level is below 6%. Under cold storage conditions with a temperature of 12-15 ºC and a humidity of 35-45%, seeds can be stored for 3-4 years provided that the initial seed moisture is maintained at 6% level.

Fig. 13. Drying of onion umbels in threshing yard.

Seed Standards

Some of the criteria are prescribed by Indian Minimum Seed Certification Standards for onion seed production and for seed certification need for the quality seeds.

Land Requirements

Land selected for onion seed production should free from volunteer plants.

Field Inspection

i. Bulb production stage - Minimum of Two Inspections

1. After transplanting, the initial inspection should be made to determine isolation distance, volunteer plants, off-types, including bolters, and other relevant factors.
2. After the bulbs have been harvested, the second inspection will be performed to verify their true characteristics.

ii. Seed production stage - Minimum of Four Inspections

1. Before flowering, the first inspection should be carried out to determine isolation distance, volunteer plants, off-types, including bottlers, and other relevant factors.
2. The second and third inspections shall be conducted during flowering to check isolation, off-types, and other relevant factors.
3. At maturity, the fourth inspection will be conducted to confirm the varietal true nature and other relevant factors.

Field Standards

Isolation

Onion seed fields shall be isolated from the contaminants shown in the below table (Table2) by the distances specified in columns, isolation distance for seed production is kept between two varieties or two plots of the same variety.

General Requirements

Table 2. General isolation requirements

Isolation	Minimum distance (meters)			
Requirements	Bulb production stage		Seed production stage	
Class	Foundation	Certified	Foundation	Certified
Fields of other varieties	5	5	1000	500
Fields of the same variety	5	5	1000	500

Specific Requirements

Table 3. Specific purity requirements

Specific Requirements	**Foundation**	**Certified**
*Bulbs not conforming to the varietal characteristics	0.10%	0.20%
**Off types	0.10%	0.20%

*Maximum permitted at second inspection at mother bulb production stage
**Maximum permitted at and after flowering at seed production stage

Seed Standards

Table 4. Minimum and maximum limits of particulars for seed certification

Class	**Foundation**	**Certified**
Pure seed (minimum)	98.00%	98.00%
Inert matter (maximum)	2.00%	2.00%
Other crop seeds (maximum)	5/kg	10/kg
Weed seeds (maximum)	5/kg	10/kg
Germination (minimum)	70%	70%
Moisture (maximum)	8.00%	8.00%
Vapour-proof containers (maximum)	6.00%	6.00%

References

Cattivelli, A., Conte, A., Martini, S., Tagliazucchi, D. 2021. Influence of Cooking Methods on Onion Phenolic Compounds Bioaccessibility. Foods. 10(5):1023.

Delaplane, K. S. and Mayer, D.F. (2000). Crop Pollination by Bees. CABI Publishing. New York, USA.

FAOSTAT. 2020. Onion production, area, and productivity. Available at: http:// faostat3.fao. org/browse/Q/QC/E (Accessed October 06, 2022).

Gnanasundari, K., Rama Krishna, K., Srivignesh, S., Manish Kumar, Harish, A. and Ramesh Kumar. A. (2023). Processing and Value Addition in Onion. Planta, (6): 1097– 1103.

Kiani, Z., Mashayekhi, K., Golubkina, N., Mousavizadeh, S.J., Nezhad, K.Z., Caruso, G. 2023. Agronomic, physiological, genetic and phytochemical characteristics of onion varieties influenced by daylength requirements. Agriculture, (13): 697.

Lawande, K.E. and Murkute, A.A. (2011) Initiatives in onion and garlic cool chain management. Proceedings of National Symposium on Postharvest Packaging, Cold-chain Logistics and Instrumentation Techniques for Quality and Safety of Perishables. December 19 – 20, 2011. Central Institute of Post-Harvest Engineering and Technology, Ludhiana, India. pp. 183.

Mahajan, V., Gupta, A.J., Thangasamy, A., Gawande, S.J. and Singh, M. 2017. Quality seed production of onion. Indian Horticulture, 62(6).

Murkute, A.A. (2012) Evaluation of onion varieties in ambient storage. DOGR News 16, 2-3.

Pareek, S., Sagar, N.A., Sharma, S., Kumar, V. Onion (Allium cepa L.). Chemistry and Human Health. In: Fruit and Vegetable Phytochemicals, 2nd ed.; Yahia, E.M., Ed.; John Wiley & Sons: Chichester, UK; Hoboken, NJ, USA, 2017; pp. 1145–1161.

Pike, L. M. 1986. Onion breeding. In: Bassett M.J. (ed), Breeding vegetable crops. The AVI publishing company, Inc., Connecticut, pp: 357-394.

Ray, P. and Gupta, H. N. 1980. "Charaka Samhita" Scientific synopsis. Indian National Science Academy, New Delhi

Selvaraj, S. 1976. Onion: queen of the kitchen. Kisan World,3(12):32–34.

www.dogr.gov.in;

www.icar.org.in

www.iihr.gov.in;

Index